W0091888

Hydraulic Fracturing Chemicals and Fluids Technology

Hydraulic Fracturing Chemicals and Fluids Technology

Editor

Narendra Parihar

Hydraulic Fracturing Chemicals and Fluids Technology
Edited by **Narendra Parihar**

Printed in 2017

ISBN: 978-1-68117-401-3

Library of Congress Control Number: 2015941592

SCITUS Academics LLC,
616, Corporate Way, Suite 2, 4766,
Valley Cottage, NY 10989

www.scitusacademics.com

Notice

Reasonable efforts have been made to publish reliable data and views articulated in the chapters are those of the individual contributors, and not necessarily those of the editors or publishers. Editors or publishers are not responsible for the accuracy of the information in the published chapters or consequences of their use. The publisher believes no responsibility for any damage or grievance to the persons or property arising out of the use of any materials, instructions, methods or thoughts in the book. The editors and the publisher have attempted to trace the copyright holders of all material reproduced in this publication and apologize to copyright holders if permission has not been obtained. If any copyright holder has not been acknowledged, please write to us so we may rectify.

Contents

Preface

Hydraulic Fracturing Chemicals and Fluids Technology provides an easy-to-use manual to create fluid formulations that will meet project-specific needs while protecting the environment and the life of the well. Fink creates a concise and comprehensive reference that enables the engineer to logically select and use the appropriate chemicals on any hydraulic fracturing job. The book devoted entirely to hydraulic fracturing chemicals, Fink eliminates the guesswork so the engineer can select the best chemicals needed on the job while providing the best protection for the well, workers and environment. When classifying fracturing fluids and their additives, it is important that production, operation, and completion engineers understand which chemical should be utilized in different well environments.

Editor

Chapter 1

Hydraulic Fracturing and its Peculiarities

Stefano Secchi[1] and Bernhard A Schrefler[2]

[1]Institute of system science ISIB – CNR, corso Stati Uniti 4, Padua 35127, Italy

[2]Department of Civil, Environmental and Architectural Engineering, University of Padua, 9 via Marzolo, Padua 35123, Italy

ABSTRACT

Background

Simulation of pressure-induced fracture in two-dimensional (2D) and three-dimensional (3D) fully saturated porous media is presented together with some peculiar features.

Methods

A cohesive fracture model is adopted together with a discrete crack and without predetermined fracture path. The fracture is filled with interface elements which in the 2D case are quadrangular and triangular elements and in the 3D case are either tetrahedral or wedge elements. The Rankine criterion is used for fracture nucleation and advancement. In a 2D setting the fracture follows directly the direction normal to the maximum principal stress while in the 3D case the fracture follows the face of the element around the fracture tip closest to the normal direction of the maximum principal stress at the tip. The procedure requires continuous updating of the mesh around the crack tip to take into account the evolving geometry. The updated mesh is obtained by means of an efficient mesh generator based on Delaunay tessellation. The governing equations are written in the framework of porous media mechanics and are solved numerically in a fully coupled manner.

Results

Numerical examples dealing with well injection (constant inflow) in a geological setting and hydraulic fracture in 2D and 3D concrete dams (increasing pressure) conclude the paper. A counter-example involving thermomecanically driven fracture, also a coupled problem, is included as well.

Conclusions

The examples highlight some peculiar features of hydraulic fracture propagation. In particular the adopted method is able to capture the hints of Self-Organized Criticality featured by hydraulic fracturing.

BACKGROUND

Fluid-driven fracture propagating in porous media is widely used in geomechanics to improve the permeability of reservoirs in oil and gas recovery or of geothermal wells. Another application of importance is related to the overtopping stability analysis of dams. In the case of reservoir engineering, water is forced under high pressure deep into the ground by injection into a well. The fluid, usually mixed with sand and some chemicals, penetrates in the reservoir rock, opening long cracks (fracking). Horizontal drilling together with hydraulic fracturing makes the extraction of tightly bound natural gas from shale formations economically feasible [1]. In the field, it is unfortunately rather difficult to obtain direct information about the evolution of the crack in the ground, and very little data are known or accessible. Two types of measurements are mainly performed: monitoring of pressure fluctuations at the injection pump and registration of acoustic emissions at the surface [2]. Fracking can also induce small earthquakes [3]. Pressure-induced fracture propagation presents some peculiar features such as pressure peaks and stepwise advancement, which have been discovered only recently and need further investigation. It is recalled that differently from tensile experiments where the crack surfaces are stress free, in hydraulic fracturing, these surfaces are loaded by a pressure distribution resulting from the invading fluid or gas [2]. Simulation is an extremely useful tool to obtain more insight into the problem. The paper addresses this issue.

Contributions to the mathematical modelling of fluid-driven fractures have been made continuously since the 1960s, beginning with Perkins and Kern [4]. These authors made various simplifying assumptions, for instance, regarding fluid flow, fracture shape and velocity leakage from the fracture. For other analytical solutions in the frame of linear fracture mechanics, assuming the problem to be stationary, see [5-9]. They suffer the limits of an analytical approach, in particular the inability to represent an evolutionary problem in a domain with a real complexity. An analysis of solid and fluid behaviour near the crack tip can be found in [10,11]. Boone and Ingraffea [12] present a numerical model in the context of linear fracture mechanics which allows for fluid leakage in the medium surrounding the fracture and assumes a moving crack depending on the applied loads and material properties. Tzschichholz and

Herrmann [2] used a two-dimensional (2D) lattice model for constant injection rate and homogeneous and heterogeneous material which only breaks under tension. Carter et al. [13] show a fully three-dimensional (3D) hydraulic fracture model which neglects the fluid continuity equation in the medium surrounding the fracture. A discrete fracture approach with remeshing in an unstructured mesh and automatic mesh refinement is used by Schrefler et al. [14]. An element threshold number (number of elements over the cohesive zone) was identified to obtain mesh-independent results. This approach has been extended to 3D situation in [15]. Extended finite elements (XFEM) have been applied to hydraulic fracturing in a partially saturated porous medium by Réthoré et al. [16] in a 2D setting. In this case, a two-scale model has been developed for the fluid flow: in the cohesive crack, Darcy's equation is used for flow in a porous medium, and identities are derived that couple the local momentum and mass balances to the governing equations for the unsaturated medium at macroscopic level. As an example, the rupture of a saturated square plate (0.25×0.25 m) in plane strain conditions is investigated under a prescribed fixed vertical velocity $v = 2.35 \times 10^{-2}$ μm/s in the opposite direction at the top and bottom of the plate (tensile loading). The mesh used consists of 20×20 quadrilateral elements (12.5×12.5 mm each) with bilinear shape functions, and the time step size is 1 s. In the cracked region, the elements are further divided in four triangles. Mohammadnejad and Khoei [17] solve the same problem also with XFEM, using full two-phase flow throughout the region. Darcy flow is assumed within the crack. Finer meshes are used as above (smallest element size 4.5×4.5 mm) and much lower time steps (0.25 to 0.125 s). Cavitation is found in both papers, also due to the impervious boundary conditions chosen. Partition of unity finite elements (PUFEM) are used for 2D mode I crack propagation in saturated ionized porous media by Kraaijeveldt et al. [18]. A pull test, a delamination test and an osmopolarity test are simulated with rather fine regular meshes (quadrangular elements with side length of the order of 2 mm and lower) and time step size down to 0.1 s. The time and space discretizations, including the element threshold number used for the solutions, are extremely important for catching the phenomena described next.

We address now the peculiar behaviour of hydraulic fracture propagation which has been observed only by a minority of the above-mentioned authors, but has been confirmed experimentally. Tzschichholz and Herrmann [2] have evidenced with their lattice model and constant injection rate a drop in pressure in time and oscillations on short time scales. These authors explain this by the fact that at the beginning high pressures are needed to push the fluid into the crack. The crack is enlarged and the pressure drops because the enlarged crack can now be opened much more easily than before. The pressure goes down although additional fluid has been added to the crack in the time step. If the pressure drops too much, the stresses at the crack tip fall below their cohesion value and the crack cannot grow at the next time step. By injecting more fluid into the crack, the pressure increases linearly in time until the cohesion forces can be overcome again. Using arguments from continuum mechanics, the authors show that the obtained value for pressure decline in the long term agrees acceptably with their numerical results. The short-term deviations are due the lattice model and the ensuing pressure drops. Oscillations are also obtained for the stored lattice energy. The breaking process is discontinuous in time with time intervals of quiescence where all beams on the crack surface are stressed below their cohesion thresholds and the acting pressure increases linearly in time. Tzschichholz and Herrmann [2] also find a temporal clustering of the breaking events, calling such a sequence bursts (avalanche behaviour). The bursts are unevenly distributed in time and occur relatively often for small times and become rarer later. There is resemblance between the obtained data and magnitude records of earthquakes or of acoustic emission records from laboratory experiments. We have shown with our porous media mechanics model in a 2D setting [14] that in the case of hydraulic fracturing the fracture advances stepwise. Two types of mesh refinement in space and refinement in time were used, but the stepwise advancement did not disappear. Such steps do not appear in other coupled solutions involving cohesive fracture, as e.g. the thermo-elastic one of[19] where the crack surfaces are stress free. The stepwise advancement and flow jumps were also found by Kraaijeveld [20] with a strong and a weak discontinuity model for flow. In [18], the stepwise advancement in mode I crack propagation is difficult to see because a continuous pressure profile across the crack is used. However, continuous pressure profile only works for sufficiently fine meshes. If the mesh is sufficiently fine, then the discretization can resolve the steep pressure gradients along the crack, but the advantage of PUFEM which allows keeping the mesh pretty rough all over the continuum is lost. Hence, dealing with the stepwise progression of the crack in this mode I model is only possible with a finer mesh than the one used (JM Huyghe, personal communication). This is why the authors state that the physical phenomenon challenges the numerical scheme. In mode II, as shown in [20,21], this problem does not appear because a discontinuous pressure across the crack is accounted for. There it is not attempted to resolve the steep pressure gradient, but this gradient is reconstructed afterwards, using the Terzaghi analytical solution for pressure diffusion. This two-step procedure allows using a rough mesh and still handling a realistic pressure gradient. Stepwise crack advancement can clearly be observed in the crack length histories of Figure four of [17], while it does not appear in the solution for the same problem in [16]. The cohesive fracture length for this problem is estimated with Barenblatt's expression (see Equation 22) [22] to be about 136 mm. Hence, there are about 10 elements over the cohesive zone in [16] and 30 elements in [17]. The first value is probably below the element threshold number for this type of problem, even with XFEM (see also the large time steps used), while the second one is sufficient even for standard elements. While both use XFEM, the two-step procedure and the large time step size and coarse mesh in [16] hide the problem.

Finally, stepwise advancement and flow are also mentioned in [23], where PUFEM is used for 2D poroelastic media. Their method still suffers from mesh dependence because the crack propagates through one element each time step. Hence, their conclusions are not definite. However, Pizzocolo et al. [24] confirmed stepwise advancement experimentally with a test on a small hydrogel disk. The duration of the pause t between steps is found to be inversely related to the hydraulic permeability$K$ according to $t = \Delta x^2 / KE$ with E Young's modulus and x length of the step. A possible explanation for the stepwise behaviour observed in [20,21,24] put forward in [24] is that an incompressible fluid consolidation comes into play which prevents tip advancement until the overpressures due to the last advancement have been dissipated, and the stress has been transferred again to the solid phase. This implies the existence of pressure peaks after each advancement stage. During the period of quiescence, the effective stress is below the breaking threshold. Consolidation as a possible explanation for the stepwise advancement needs further investigation in the case of fluid injection, because for some permeability values the tip pressure goes down to zero as shown below on an example. The existence of periods of quiescence is in line with the findings of [2]. We will show that this phenomenon is not only relevant for small structures, where it has been observed experimentally, but also for large structures such as underground soil masses and dams. In that case, the fracture length is much larger, but the phenomenon is still there and the bursts can be felt at great distances compared to the crack length.

METHODS

The first subsection presents the fracture model, the second subsection summarizes the governing equations and their numerical solution by means of the finite element method and the third subsection explains the adopted fracture advancement procedure and the required refinements necessary to obtain mesh-independent results.

The Fracture Model

We use a discrete crack model for a situation depicted in Figure 1: Ω is the domain, Γ_e is the boundary of the fully saturated porous material surrounding the crack, Γ′ is the crack boundary and $\bar{\Omega}$is the domain inside the crack filled with fluid only. There is fluid exchange between the crack and the surrounding porous medium. The mechanical behaviour of the solid phase at a distance from the process zone is assumed to obey a Green elastic or hyperelastic material behaviour [14].

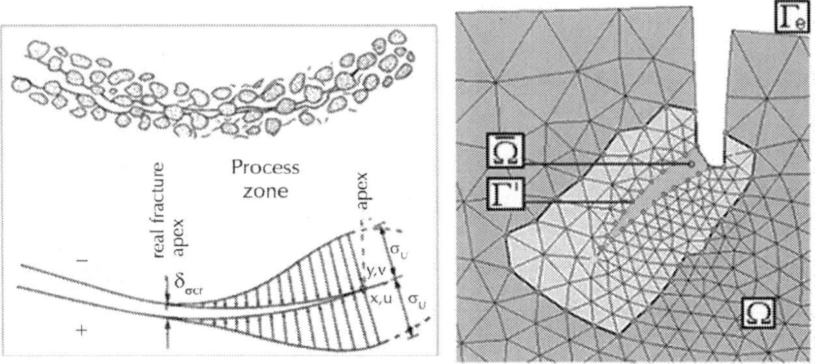

Figure 1: Hydraulic fracture domain and cohesive crack geometry.Definition of the hydraulic fracture domain, reprinted from [15], Copyright (2012), with permission from Springer, and the cohesive crack geometry.

For the fracture itself, we use a cohesive fracture model. Between the real fracture apex which appears at macroscopic level and the apex of a fictitious fracture, there is a process zone where cohesive forces act (see Figure 1). Following [22,25,26], the cohesive law for mode I crack opening with monotonically increasing opening is

$$\sigma = \sigma_0\left(1 - \frac{\delta_\sigma}{\delta_{\sigma cr}}\right) \tag{1}$$

σ_0 being the maximum cohesive traction (closed crack), δ_σ the current relative displacement normal to the crack, $\delta_{\sigma cr}$ the maximum opening with exchange of cohesive tractions and $G_c = \sigma_0 \times \delta_{\sigma cr} / 2$ the fracture energy. If after some opening $\delta_{\sigma 1} < \delta_{\sigma cr}$, the crack begins to close and tractions obey a linear unloading as

$$\sigma = \sigma_0 \left(1 - \frac{\delta_{\sigma 1}}{\delta_{\sigma cr}}\right) \frac{\delta_\sigma}{\delta_{\sigma 1}} \tag{2}$$

When the crack reopens, Equation 2 is reversed until the opening $\delta\sigma_1$ is recovered; then, tractions obey again Equation 1.

When tangential relative displacements of the sides of a fracture in the process zone cannot be disregarded, mixed mode crack opening takes place. This is often the case of a crack moving along an interface separating two solid components. In fact, whereas the crack path in a homogeneous medium is governed by the principal stress direction, the interface has an orientation that is usually different from the principal stress direction. The mixed cohesive mechanical model involves the simultaneous activation of normal and tangential displacement discontinuity and corresponding tractions. For the pure mode II, the relationship between tangential tractions and displacements is

$$\tau = \tau_0 \frac{\delta_\sigma}{\delta_{\sigma cr}} \frac{\delta_\tau}{|\delta_\tau|} \tag{3}$$

τ_0 being the maximum tangential stress (closed crack), δ_τ the relative displacement parallel to the crack and $\delta_{\sigma cr}$ the limiting value opening for stress transmission. The unloading/loading from/to some opening $\delta_{\sigma 1} < \delta_{\sigma cr}$ follows the same behaviour as for mode I.

For the mixed mode crack propagation, the interaction between the two cohesive mechanisms is treated as in [27]. By defining an equivalent or effective opening displacement δ and the scalar effective traction t as

$$\delta = \sqrt{\beta^2 \delta_\tau^2 \delta_\sigma^2}, \quad t = \sqrt{\beta^{-2} \tau^2 + \sigma^2} \tag{4}$$

the resulting cohesive law is

$$\mathbf{t} = \frac{t}{\delta} \left(\beta^2 \boldsymbol{\delta}_\tau + \boldsymbol{\delta}_\sigma\right) \tag{5}$$

β being a suitable material parameter that defines the ratio between the shear and the normal critical components. For more details, see [14].

Governing Equations and Their Discretization in Space and Time

Taking into account the cohesive forces and the symbols of Figure 1, the linear momentum balance of the mixture, discretized in space with finite elements according to the standard Galerkin procedure[28] is written as

$$\mathbf{M}\dot{\mathbf{v}} + \int_{\Omega} \mathbf{B}^T \boldsymbol{\sigma}'' d\Omega - \mathbf{Q}\mathbf{p} - \mathbf{f}^{(1)} - \int_{\Gamma'} (\mathbf{N}^u)^T \mathbf{c}\, d\Gamma' = \mathbf{0} \tag{6}$$

where c is the cohesive traction on the process zone as defined above.

The fully saturated medium surrounding the fracture has constant absolute permeability, while for the permeability within the crack, the Poiseuille or cubic law is assumed. This permeability does not depend on the rock type or stress history but is defined by crack aperture only. Deviation from the ideal parallel surface conditions causes only an apparent reduction in flow and can be incorporated into the cubic law, which reads as [29]

$$k_{ij} = \frac{1}{f}\,\frac{w^2}{12} \tag{7}$$

w being the fracture aperture and f a coefficient in the range 1.04 to 1.65 depending on the solid material. In the following, this parameter will be assumed as constant and equal to 1.0. Incorporating the Poiseuille law into the weak form of water mass balance equation within the crack and discretizing in space by means of the finite element method results in

$$\tilde{\mathbf{H}}\mathbf{p} + \tilde{\mathbf{S}}\dot{\mathbf{p}} + \int_{\Gamma'} (\mathbf{N}^p)^T \mathbf{q}^w\, d\Gamma' = \mathbf{0} \tag{8}$$

With

$$\tilde{\mathbf{H}} = \int_{\bar{\Omega}} (\nabla \mathbf{N}^p)^T \frac{w^2}{12\mu_w} \nabla \mathbf{N}^p \, d\bar{\Omega} \tag{9}$$

$$\tilde{\mathbf{S}} = \int_{\bar{\Omega}} (\mathbf{N}^p)^T \frac{1}{K_f} \mathbf{N}^p \, d\bar{\Omega} \tag{10}$$

μ_w is the dynamic viscosity and K_f the bulk modulus of the fluid. The last term of (8) represents the leakage flux into the surrounding porous medium across the fracture borders and is of paramount importance in hydraulic fracturing techniques. This term can be represented by means of Darcy's law using the medium permeability and pressure gradient generated by the application of water pressure on the fracture lips. No particular simplifying hypotheses are hence necessary for this term. This equation can be directly assembled at the same stage as the mass balance Equation 11 for the saturated medium surrounding the crack, because both have the same structure: only the parameters have to be changed in the appropriate elements depending whether they belong to the fracture or to the surrounding medium.

The discretized mass balance equation for the porous medium surrounding the fracture is

$$\mathbf{Q}^T \dot{\mathbf{u}} + \mathbf{H}\mathbf{p} + \mathbf{S}\dot{\mathbf{p}} - \mathbf{f}^{(2)} - \int_{\Gamma} (\mathbf{N}^p)^T \mathbf{q}^w d\Gamma' = \mathbf{0} \tag{11}$$

where q^w represents the water leakage flux along the fracture toward the surrounding medium of Equation 7. This term is defined along the entire fracture, i.e. the open part and the process zone. It is worth mentioning that the topology of the domains Ω and $\bar{\Omega}$ changes with the evolution of the fracture. In particular, the fracture path, the position of the process zone and the cohesive forces are unknown and must be regarded as products of the mechanical analysis.

Discretization in time is then performed with time discontinuous Galerkin approximation following[30,31]. Denoting with $I_n = \left(t_n^-, t_{n+1}^+\right)$ a typical incremental time step of size $\Delta t = t_{n+1} - t_n$, the weighted residual forms are

$$\int_{I_n} \delta\mathbf{v}^T \left(\mathbf{M}\dot{\mathbf{v}} + \mathbf{K}\mathbf{u} - \mathbf{Q}\mathbf{p} - \mathbf{f}^{(1)}\right)\mathrm{dt} + \int_{I_n} \delta\mathbf{u}^T \mathbf{K}(\dot{\mathbf{u}} - \mathbf{v})\mathrm{dt} + \\ + \delta\mathbf{u}^T\big|_{t_n} \mathbf{K}\left(\mathbf{u}_n^+ - \mathbf{u}_n^-\right)\mathrm{dt} + \delta\mathbf{v}^T\big|_{t_n} \mathbf{M}\left(\mathbf{v}_n^+ - \mathbf{v}_n^-\right) = \mathbf{0} \tag{12}$$

$$\int_{I_n} \delta\mathbf{p}^T \left(\mathbf{Q}^T\mathbf{v} + \mathbf{S}\mathbf{s} + \mathbf{H}\mathbf{p} - \mathbf{f}^{(2)}\right)\mathrm{dt} + \int_{I_n} \delta\mathbf{p}^T \mathbf{S}(\dot{\mathbf{p}} - \mathbf{s})\mathrm{dt} + \\ + \delta\mathbf{p}^T\big|_{t_n} \mathbf{S}\left(\mathbf{p}_n^+ - \mathbf{p}_n^-\right)\mathrm{dt} = \mathbf{0} \tag{13}$$

with the constraint conditions

$$\begin{aligned} \dot{\mathbf{u}} - \mathbf{v} &= \mathbf{0} \\ \dot{\mathbf{p}} - \mathbf{s} &= \mathbf{0} \end{aligned} \tag{14}$$

Subscripts -/+ indicate quantities immediately before and after the generic time station. Field variables and their first time derivatives at time $t \in [t_n, t_{n+1}]$ are interpolated by linear time shape functions, and the following discretized equations are obtained

$$\begin{aligned} \mathbf{u}_n &= \mathbf{u}_n^- + \frac{\Delta t}{2}\left(\mathbf{v}_n^+ - \mathbf{v}_{n+1}^-\right) \\ \mathbf{u}_{n+1} &= \mathbf{u}_n^- + \frac{\Delta t}{2}\left(\mathbf{v}_n^+ + \mathbf{v}_{n+1}^-\right) \\ \mathbf{s}_n &= \frac{1}{\Delta t}\left(\mathbf{p}_{n+1} + 3\mathbf{p}_n - 4\mathbf{p}_n^-\right) \\ \mathbf{s}_{n+1} &= \frac{1}{\Delta t}\left(\mathbf{p}_{n+1} + 3\mathbf{p}_n + 2\mathbf{p}_n^-\right) \end{aligned} \tag{15}$$

$$\left(\frac{1}{2}\mathbf{M}-\frac{5}{36}\Delta t^2\mathbf{K}\right)\mathbf{v}_n+\left(\frac{1}{2}\mathbf{M}+\frac{1}{36}\Delta t^2\mathbf{K}\right)\mathbf{v}_{n+1}+\frac{\Delta t}{3}\mathbf{Q}\mathbf{p}_n+$$
$$+\frac{\Delta t}{6}\mathbf{Q}\mathbf{p}_{n+1}=-\frac{\Delta t}{2}\mathbf{K}\mathbf{u}_n^-+\mathbf{M}\mathbf{v}_n^-+\int_{I_n}N_1(t)\mathbf{f}^{(1)}\,\mathrm{dt}$$
$$\left(-\frac{1}{2}\mathbf{M}-\frac{7}{36}\Delta t^2\mathbf{K}\right)\mathbf{v}_n+\left(\frac{1}{2}\mathbf{M}+\frac{5}{36}\Delta t^2\mathbf{K}\right)\mathbf{v}_{n+1}+\frac{\Delta t}{3}\mathbf{Q}\mathbf{p}_n+$$
$$+\frac{\Delta t}{3}\mathbf{Q}\mathbf{p}_{n+1}=-\frac{\Delta t}{2}\mathbf{K}\mathbf{u}_n^-+\int_{I_n}N_2(t)\mathbf{f}^{(1)}\,\mathrm{dt}$$
$$\frac{\Delta t}{3}\mathbf{Q}^T\mathbf{v}_n\frac{\Delta t}{6}\mathbf{Q}^T\mathbf{v}_{n+1}+\left(\frac{1}{2}\mathbf{S}+\frac{\Delta t}{3}\mathbf{H}\right)\mathbf{p}_n+\left(\frac{1}{2}\mathbf{S}+\frac{\Delta t}{6}\mathbf{H}\right)\mathbf{p}_{n+1}=$$
$$=\mathbf{S}\mathbf{p}_n^-+\int_{I_n}N_1(t)\mathbf{f}^{(2)}\,\mathrm{dt}$$
$$\frac{\Delta t}{6}\mathbf{Q}^T\mathbf{v}_n\frac{\Delta t}{3}\mathbf{Q}^T\mathbf{v}_{n+1}+\left(-\frac{1}{2}\mathbf{S}+\Delta t\mathbf{H}\right)\mathbf{p}_n+\left(\frac{1}{2}\mathbf{S}+\frac{\Delta t}{3}\mathbf{H}\right)\mathbf{p}_{n+1}=\int_{I_n}N_1(t)\mathbf{f}^{(2)}\,\mathrm{dt} \tag{16}$$

The nodal displacement, velocity and pressure, n_n^-, v_n^- and p_n^-, respectively, for the current step coincide with the unknowns at the end of the previous one, hence are known in the time marching scheme and coincide with the initial condition for the first time step. The system of algebraic equations is solved with a monolithic approach using an optimized non-symmetric sparse matrix algorithm. The number of unknowns is doubled with respect to the traditional trapezoidal method.

In a quasistatic situation, adopted for the examples, the submatrices of the above equations are the usual ones of soil consolidation [28], except for

$$\dot{\mathbf{f}}^{(1)}=\int_{\Omega}\left(\mathbf{N}^u\right)^T\rho\dot{\mathbf{b}}\,d\Omega+\int_{\Gamma_t}\left(\mathbf{N}^u\right)^T\dot{\mathbf{t}}\,d\Gamma+\int_{\Gamma'}\left(\mathbf{N}^u\right)^T\dot{\mathbf{c}}\,d\Gamma' \tag{17}$$

where $\dot{\mathbf{c}}$ is the cohesive traction rate and is different from zero only if the element has a side on the lips of the fracture Γ'. Given that the liquid phase is continuous over the whole domain, leakage flux along the opened fracture lips is accounted for through the H matrix together with the flux along the crack. Finite elements are in fact present along the crack (not shown in Figure 1), which account only for the pressure field and have no mechanical stiffness. In the present formulation, non-linear terms arise through cohesive forces in the process zone and permeability along the fracture.

Fracture Advancement and Refinement Strategy

Because of the continuous variation of the domain as a consequence of the propagation of the cracks, also the boundary Γ′ and the related mechanical conditions change. Along the formed crack edges and in the process zone, boundary conditions are the direct result of the field equations, while the mechanical parameters have to be updated. The following remeshing techniques account for all these changes [15,32].

For the fracture nucleation and advancement, the Rankine criterion is used. More than one crack can open and fractures can also branch. Fracture forms and advances if the maximum principal stress exceeds in a point the fixed threshold. The fracture advancement procedure differs for 2D and 3D situations: in 2D, the fracture follows directly the direction normal to the maximum principal stress, while in 3D, the fracture follows the face of the element around the fracture tip which is closest to the normal direction of the maximum principal stress. In this last case, the fracture tip becomes a curve in space (front). The advancement of a fracture creates new nodes: in 2D, the resulting new elements for the filler at the front are triangles, while in 3D situation, they are tetrahedral. If an internal node along the process zone advances in a 3D setting, a new wedge element results in the filler [15].

At each time station *tn*, *j* successive tip (front) advancements are possible until the Rankine criterion is satisfied (Figure 2). Their number in general depends on the chosen time step increment Δt, the adopted crack length increment Δs and the variation of the applied loads. This requires continuous remeshing with a consequent transfer of nodal vectors from the old to the continuously updated mesh by a suitable operator $v_m(\Omega_{m+1}) = \aleph(v_m(\Omega_m))$. For momentum and energy conservation, the solution is repeated with the quantities of mesh m but re-calculated on the new mesh $m+1$ before advancing the crack tip [33].

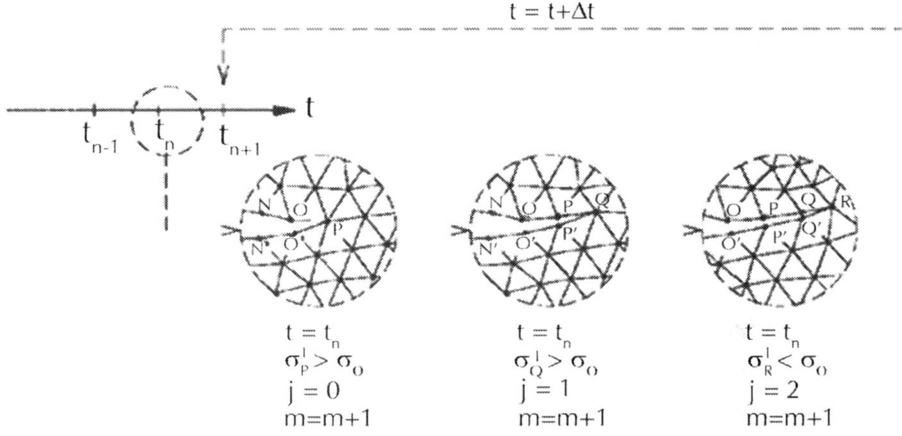

Figure 2: Multiple advancing fracture step at the same time station.

Three types of refinement are needed to obtain satisfactory results: the refinement in space in general, the satisfaction of an element threshold number over the process zone and a refinement in time. For refinement and de-refinement in space, the Zienkiewicz-Zhu error estimator is used [34]. Fluid lag, i.e. negative fluid pressures at the crack tip, may arise in the case of injection if the speed at which the crack tip advances is sufficiently high so that for a given permeability water cannot flow in fast enough to fill the created space. This as well as mesh-independent results can be obtained numerically only if an *element threshold number* is satisfied over the process zone. This number is given by the ratio of elements over the process zone and its length and can be estimated in advance from the problem at hand and the expected process zone. The number of elements over the process zone is of paramount importance and has not received sufficient attention by many authors. It is a sort of object-oriented refinement and is extensively dealt with in [14]. Adaptivity in time is obtained by means of the adopted discontinous Galerkin method in the time domain (DGT)

[33]. The error of the time-integration procedure can be defined through the jump of the solution

$$\begin{aligned}\langle \mathbf{u}_n \rangle &= \mathbf{u}_n - \mathbf{u}_n^- \\ \langle \mathbf{v}_n \rangle &= \mathbf{v}_n - \mathbf{v}_n^- \\ \langle \mathbf{p}_n \rangle &= \mathbf{p}_n - \mathbf{p}_n^- \end{aligned} \tag{18}$$

at each time station, i.e. the difference between the final point of time step n - 1 and the first point of time step n. By adopting the total energy norms as error measure, we define the following terms:

$$\begin{aligned}\|\mathbf{e}_u\|_n &= \left(\langle \mathbf{v}_n \rangle^T \mathbf{M} \langle \mathbf{v}_n \rangle + \langle \mathbf{u}_n \rangle^T \mathbf{K} \langle \mathbf{u}_n \rangle\right)^{1/2} \\ \|\mathbf{e}_{u,p}\|_n &= \left(\langle \mathbf{u}_n \rangle^T \mathbf{Q} \langle \mathbf{p}_n \rangle\right)^{1/2} \\ \|\mathbf{e}_p\|_n &= \left(\langle \mathbf{p}_n \rangle^T \mathbf{Q}^T \langle \mathbf{u}_n \rangle + \langle \mathbf{p}_n \rangle^T \mathbf{H} \langle \mathbf{p}_n \rangle \Delta t + \langle \mathbf{p}_n \rangle^T \mathbf{P}^T \langle \mathbf{p}_n \rangle\right)^{1/2} \\ \|\mathbf{e}\|_n &= \max\left\{ \|\mathbf{e}_u\|_n, \|\mathbf{e}_{u,p}\|_n, \|\mathbf{e}_p\|_n \right\} \end{aligned} \tag{19}$$

Error measures defined in Equation 19 account at the same time for the cross effects among the different fields and the ones between space and time discretizations.

The relative error is defined as in [30]

$$\eta_n = \frac{\|\mathbf{e}\|_n}{\|\mathbf{e}\|_{\max}} \tag{20}$$

where $\|\mathbf{e}\|_{\max}$ is the maximum total energy norm

$$\|\mathbf{e}\|_{\max} = \max\left(\|\mathbf{e}\|_i\right), 0 < i < n$$

When $\eta > \eta_{toll}$, the time step Δt_n is modified and a new $\Delta t'_n < \Delta t_n$ is obtained according to

$$\Delta t_n^{'} = \left(\frac{\theta \eta_{\text{toll}}}{\eta}\right)^{1/3} \Delta t_n \tag{21}$$

where $\vartheta < 1.0$ is a safety factor. If the error is smaller than a defined value $\eta_{toll,min}$, the step is increased using a rule similar to Equation

21. As it stands, the refinements in space and time are carried out sequentially, starting with the space refinement, followed by the element threshold number and then the refinement in time. An eye is kept on the satisfaction of the discrete maximum principle [35] which states that it is not possible to refine in time below a certain limit depending on the material properties without also refining in space. A proper functional would be needed to link all the three refinements. A flow chart of the numerical procedure is given in the 'Appendix'.

RESULTS AND DISCUSSION

First, we show for comparison purposes the results of cohesive fracture propagation in a thermo-elastic medium, a coupled problem solved with the method for fracture advancement outlined above[19]. The method itself of the crack tip advancement does not introduce steps, once mesh-independent results are achieved. This was also found in static problems (not shown here) such as the three-point bending test, the four-point shear test and the case of a plate with a circular hole. The problem shown here evidences also the importance of the element threshold number, i.e. the number of elements over the process zone. A three-point bending test is performed on a bimaterial specimen subjected to a thermo-mechanical loading [36]. One part of the sample is made of aluminium 6061 and the other of polymethylmethacrylate (PMMA), bonded with methacrylate adhesive. The geometry is presented in Figure 3: the sample has a notch with a sharp tip of 1-mm width and 30-mm height shifted 3 mm from the interface in the PMMA zone. The two materials present very different Young's moduli and thermal expansion coefficients, so that, when the system is subjected to heat, stresses arise near the interface as a result of the mismatch in thermal expansion.

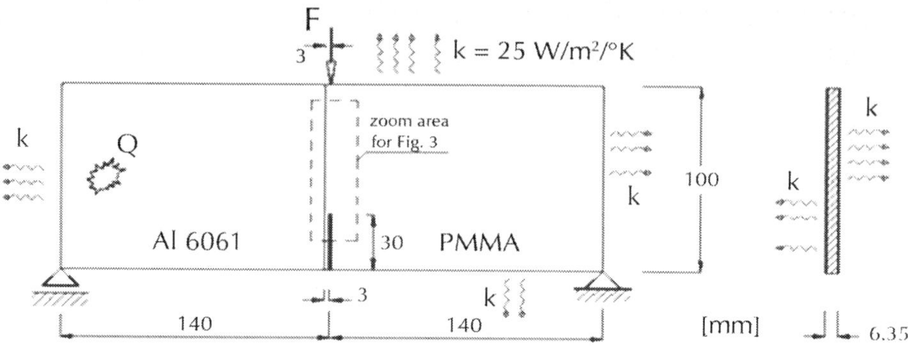

Figure 3: Geometry of the three-point bending test for a bimaterial specimen.

Two different experiments are reported in [36]. In the first, at a room temperature of 25°C, a load was applied 3 mm from the interface in the PMMA zone (Figure 3) to trigger the fracture process. The loading rate was very low and the resulting speed of crack propagation at the initial stages was also quite slow, so that quasistatic conditions can be assumed. The crack path was individuated, and stresses near the crack tip in the PMMA were measured using a shearing interferometer.

In the second experiment, the same operations were performed when the temperature of the aluminium was 60°C in steady state conditions. To reach these conditions, a cartridge heater (*Q* in Figure 3) was inserted into the aluminium part near the external vertical side. The variation in time of the PMMA temperature was checked before the fracture test, which was performed when steady state conditions were reached. The temperature of PMMA was recorded at the crack tip location, at 5 and 7 mm from the interface. Also in this case, the crack path was spotted. From the differences between the two situations, the authors gathered the thermal effects, which were independent of the magnitude of the applied mechanical load.

In the two experiments, the crack propagation trajectories differ as shown in Figure 4a,b where a zoom of the fractured specimens in correspondence of the notch is presented. In particular, the crack path is closer to the interface when the temperature is higher. The numerical results are shown in Figure 4c. The agreement is remarkable, see also [19].

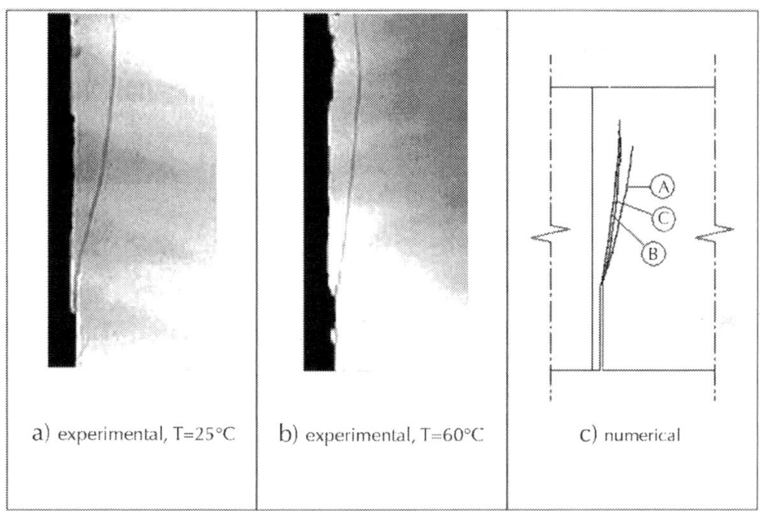

Figure 4: Zoom of the notch of the specimen with crack path trajectories. (a, b) Experimental results (reproduced from [36]). (c)Numerical results: case A, uniform temperature (25°C); case B, thermal load with E, σ_0, $\delta\sigma_{cr}$ varying with temperature; and case C, thermal load with $E = E(25°C)$, $\sigma_0 = \sigma_0(25°C)$, $\delta\sigma_{cr} = \delta\sigma_{cr}$ (25°C).

Application of Barenblatt's theory [22] for calculation of characteristic cohesive zone size l yields for PMMA

$$\ell = \frac{\pi E G_c}{8(1-\nu^2)\sigma_0^2} \cong 0.75 \text{ mm} \tag{22}$$

Our numerical results (0.8 mm) are in good accordance with this value. From this value, the choice of the crack tip advancement length can be estimated. It should be such that the heuristically determined element threshold number is satisfied (five elements in the thermo-mechanical case). Using linear elements of decreasing size, the value of the force *F* (Figure 3), corresponding to an applied vertical displacement on the same point, is calculated. Results are summarized in Figure 5. The peak of the external load and the softening branch are mesh independent once the process zone is subdivided into at least five elements with edges of 0.15 mm or smaller. This situation is handled by the mesh generator simply by locating an element source [32] at the crack tip. Its weight may be*a priori* stated and/or can be *a posteriori* updated during the adaptive remeshing procedure once the length of the process zone has been determined. What is important here is that the diagrams in this coupled problem are smooth reasonably once mesh-independent results are obtained.

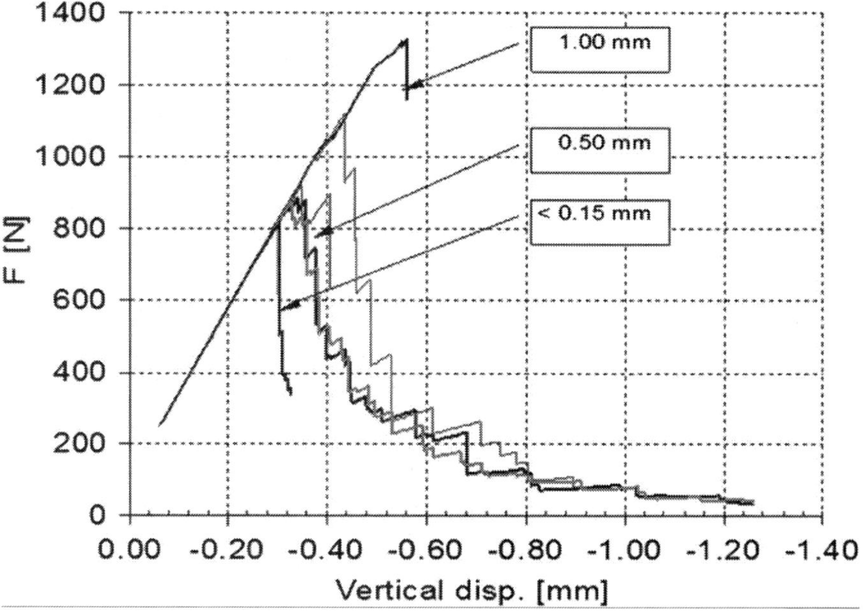

Figure 5: External force vs. vertical displacement and mesh size.

The next application deals with a hydraulically driven fracture due to fluid pumped at constant flow rate *Q* into a well in 2D conditions (plane strain) [14]. Figure 6 shows the geometry of the problem together with the initial finite element discretization. A notch with a sharp tip is present along the symmetry axis of the analyzed area.

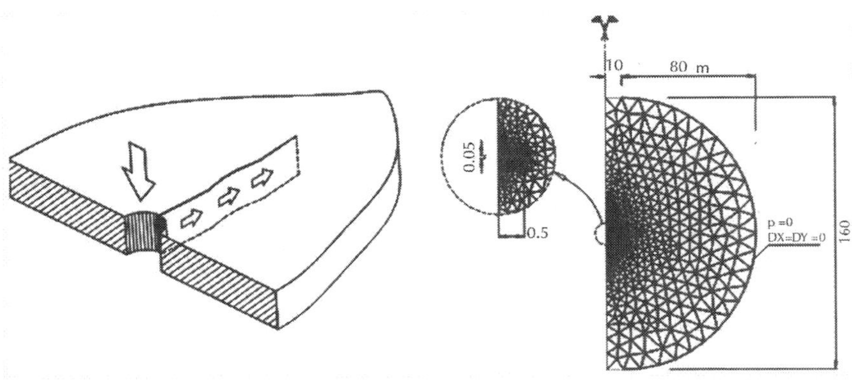

Figure 6: Problem geometry for water injection benchmark and overall discretization.

The effects of combined spatial/temporal discretizations are clearly seen in Figure 7, where the crack length is drawn versus time for different tip advancements, Δs, and time step increments, Δt. The correct time history (case E) is obtained by simultaneously reducing these two parameters, whereas the reduction of only one discretization parameter leads to errors (about ±20%) even using small tip advancement, if compared to the crack length. Again, the importance of the element threshold number is evident for the choice of Δs (the length of the process zone according to Equation 22 is 0.8 m, and about 30 elements are needed over it). It clearly appears that the crack tip velocity is very mesh sensitive. Hence, the element threshold number must be satisfied to obtain mesh-independent results.

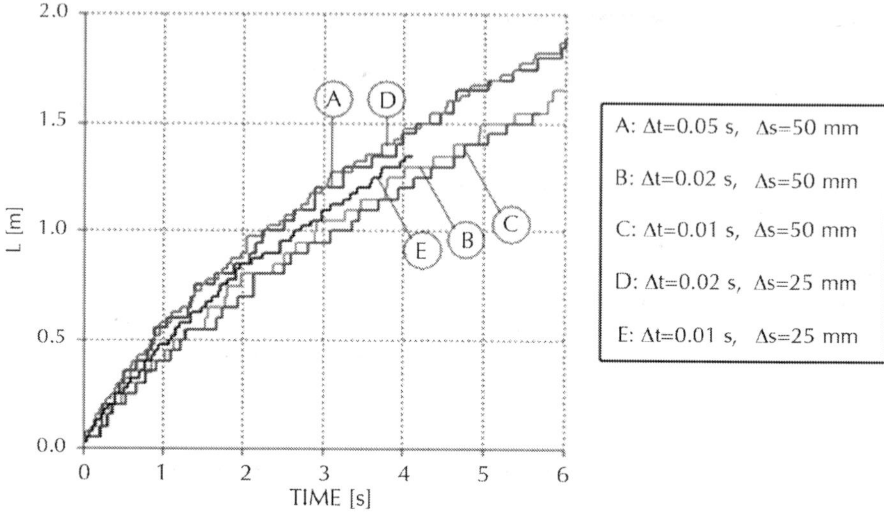

Figure 7: Crack length time history for $\boldsymbol{\mu}_w = 1 \times 10\text{-}9$ MPa s and $Q = 0.0001$ m3/s.

A lower number of elements results in wrong crack tip velocity, and the possible development of fluid lag may be missed [14]. Fluid lag corresponds to negative pressure in the process zone and determines hence different body forces (see Equation 6). The distribution of the pressure over the fracture length at time station 10 min is shown in Figure 8 for the following three combinations of dynamic viscosity and injection rate: $\mu_w = 1 \times 10^{-9}$ MPa s, $Q = 0.0001$ m^3/s; $\mu_w = 1 \times 10^{-11}$ MPa s, $Q = 0.0001$ m^3/s; and $\mu_w = 1 \times 10^{-9}$ MPa s, $Q = 0.0002$ m^3/s. The fracture length clearly varies with the chosen data. For the first combination, the pressure at the fracture tip goes almost to zero, while for lower values of μ_w, the pressure is almost constant. For high μ_w and doubled injection rate, cavitation occurs.

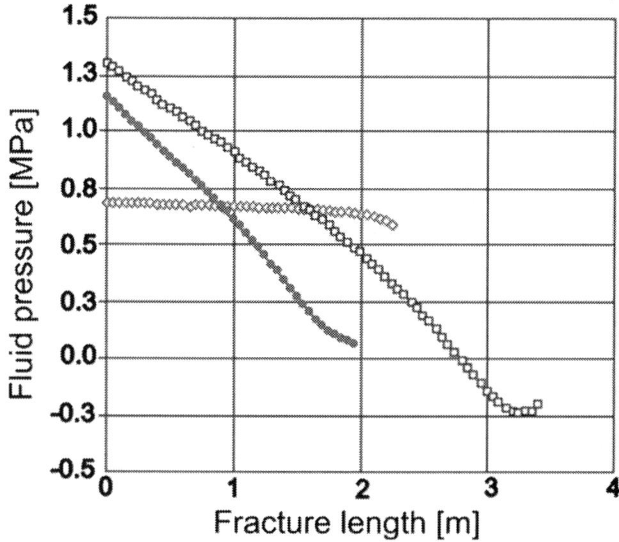

Figure 8: Distribution of the fluid pressure over the fracture length.At time station 10 min for the combinations of dynamic viscosity and injection rate: $\mu_w = 1 \times 10^{-9}$ MPa s, $Q = 0.0001$ m^3/s (red circles); $\mu_w = 1 \times 10^{-11}$ MPa s, $Q = 0.0001$ m^3/s (green diamonds); and $\mu_w = 1 \times 10^{-9}$ MPa s, $Q = 0.0002$ m^3/s (white squares).

The third case deals with the benchmark exercise A2 proposed by ICOLD [37]. The benchmark consists in the evaluation of failure conditions as a consequence of overtopping wave acting on a concrete gravity dam. Contrarily to the previous example here, we have increasing pressure. The geometry of the dam is shown in Figure 9 together with boundary conditions and an intermediate discretization. Differently from the original benchmark, the dam concrete foundation is also considered, which has been assumed homogeneous with the dam body. In such a situation, the crack path is unknown. On the contrary, when a rock foundation is present, the crack naturally develops at the interface between the dam and foundation. In Figure 9 also, the influence of the viscosity on the crack direction is evidenced (circle).

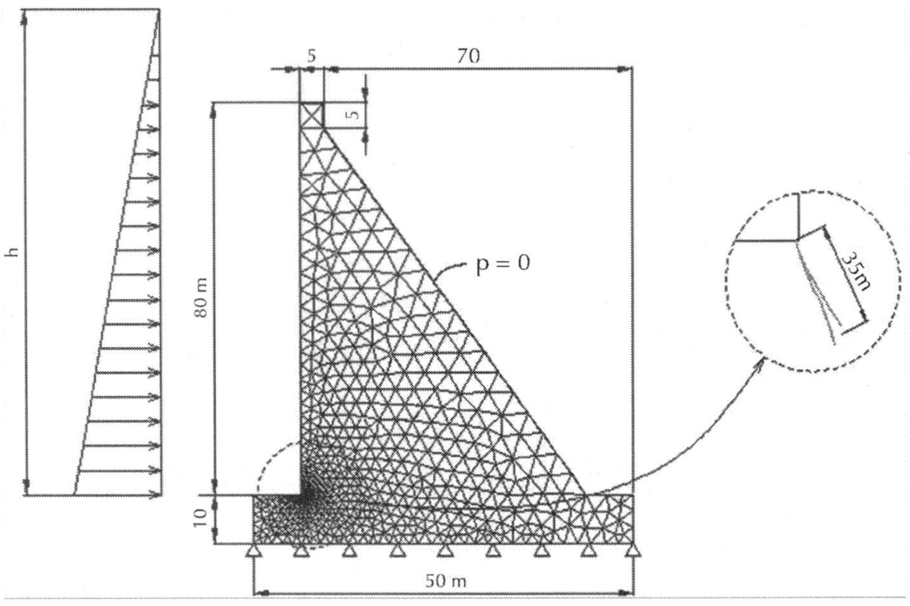

Figure 9: Problem geometry for ICOLD benchmark and calculated crack positions.

The initial condition is obtained under self-weight and the hydrostatic pressure due to water in the reservoir up to a level of 52 m. From this point, the water level increases until the overtopping level is reached (higher than the dam crest [14]). The increase of water level in the reservoir is specified in days according to the benchmark.

For an intermediate situation, the principal stress contours and the cohesive forces are shown on Figure 10. Also, fluid lag has been obtained for this situation, not shown in the picture (see [14]). The crack mouth opening displacement versus days is depicted in Figure 11 for different values of the crack tip advancement. The smallest value corresponds to the proper element threshold number. Clearly, stepwise advancement can be observed with some clustering of the steps.

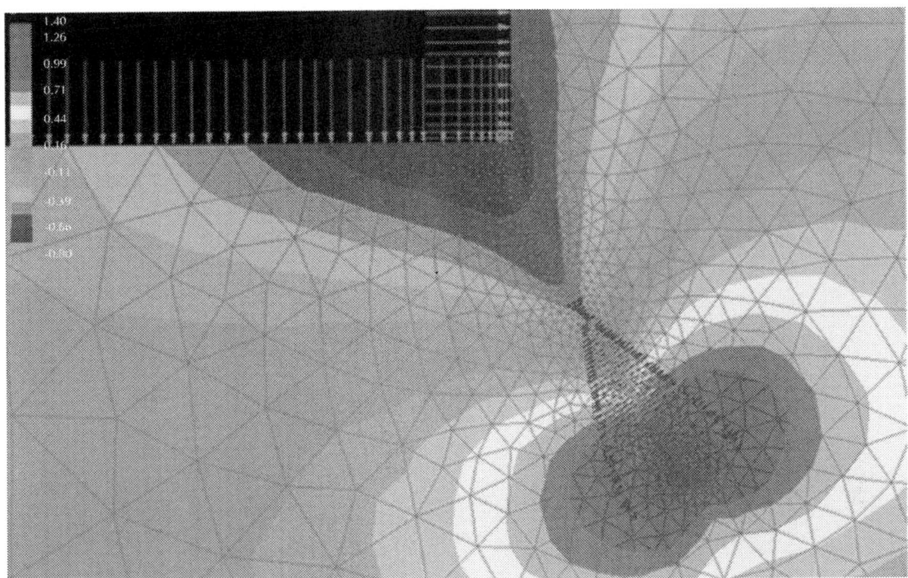

Figure 10: Zoom near the fracture on maximum principal stress contour and cohesive forces.

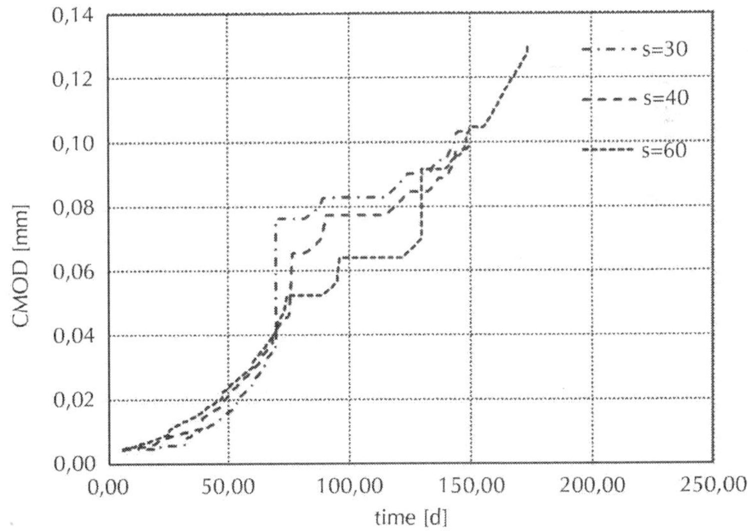

Figure 11: Crack mouth opening displacement versus time (days) for different values of the crack tip advancements (mm).

The effects of the stepwise advancement can also be felt at great distance from the actual crack: the horizontal displacements on the dam crest are effected, as can be seen from Figure 12. Only the diagram for the purely elastic solution (no crack) is smooth. Note that here the vertical scale is logarithmic and in the abscissa appear the time steps, not the actual time. This is the reason why the diagram for the elastic case is above the others.

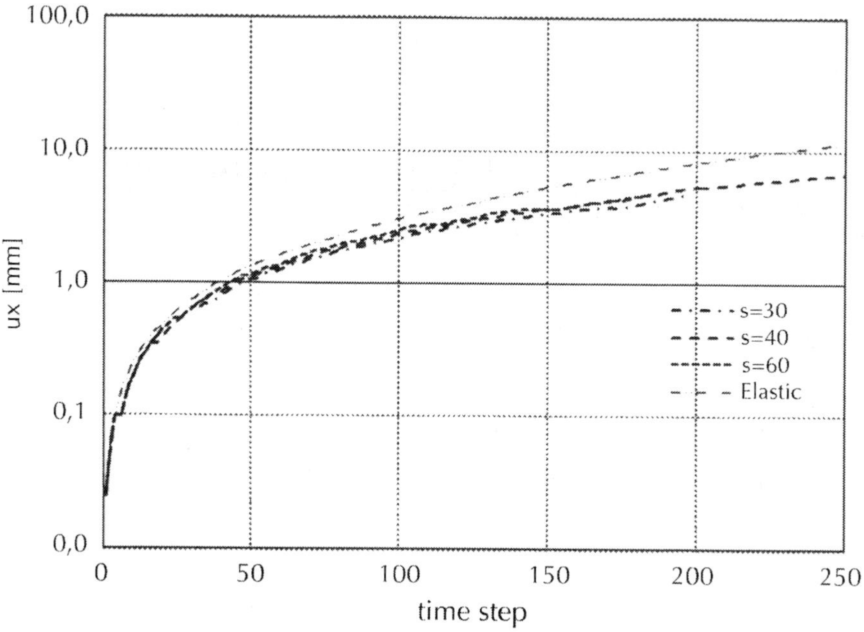

Figure 12: Horizontal displacements versus time step of the dam crest. For different values of the crack tip advancements (mm) and without fracture (elastic).

For a similar problem, a 3D solution has been obtained in [15]. In Figure 13, the mesh, the fracture, the process zone and the stress contours are shown when the fracture length is about 15 m corresponding to an intermediate step of the analysis when the water level is 80 m. The horizontal displacement of the dam crest is drawn versus time in Figure 14. The following situations are considered: no fracture at all (elastic); dry fracture (fracture), i.e. water pressure acts only on the dam, not on the crack lips; hydrostatic water pressure in the crack, constant over the crack length (hydraulic fracture); and fully coupled solution with water pressure varying over the crack length (u-*p*). The last one has fluid exchange between the crack and surroundings. The results for the last case correspond to an intermediate value between the others because the pressure is diminishing towards the crack tip, reaching even negative values there (cavitation).

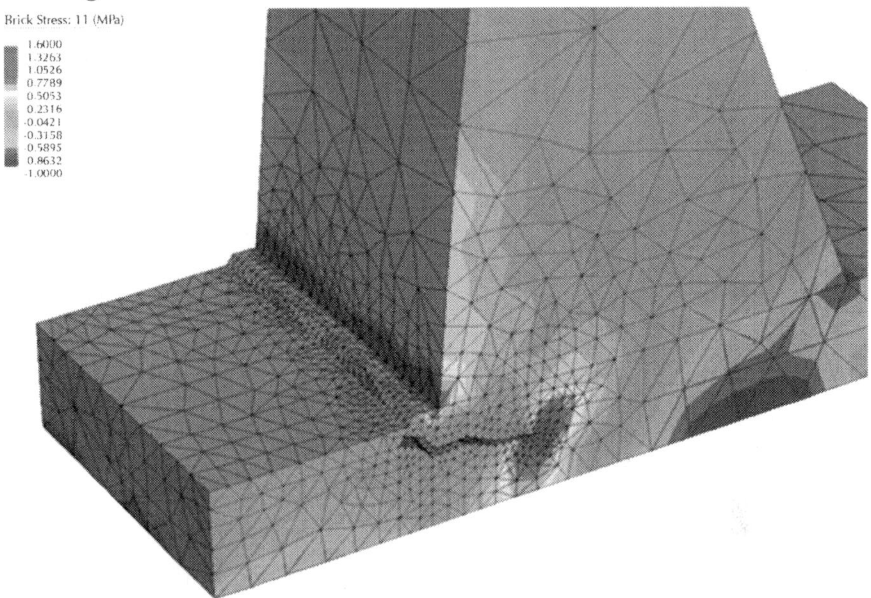

Figure 13: Mesh, fracture, process zone and stress contours. With a fracture length of about 15 m corresponding to an intermediate step of the analysis when the water level is 80 m.

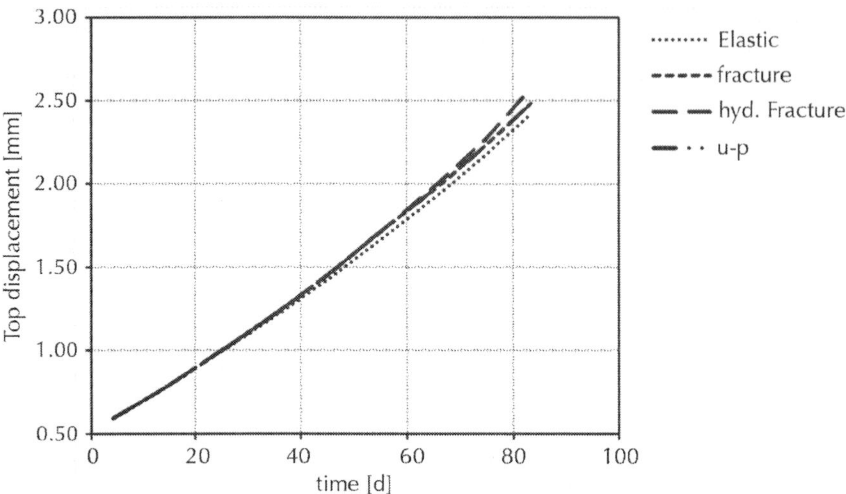

Figure 14: Horizontal displacement of the dam crest versus time. No fracture at all (elastic); dry fracture (fracture), i.e. water pressure acts only on the dam, not on the crack lips; hydrostatic water pressure in the crack, constant over the crack length (hydraulic fracture); and fully coupled solution with water pressure varying over the crack length and fluid exchange between the crack and surroundings (u-*p*).

The relative variations of the horizontal crest displacements according to

$$\|u\| = \left(\frac{u_i}{u_{\mathrm{el}}} - 1\right) \cdot 100 \tag{23}$$

with u_i referring to the studied cases and u_{el} to the elastic solution, are drawn in Figure 15. The largest steps correspond to the situations where fluid is present in the crack and may have pressure exchange (consolidation) with the material surrounding the process zone. Note that these variations are felt on the dam crest, while the pressure-induced fracturing happens on the bottom of the dam. The 3D results have however only qualitative value because the element threshold number would require finer meshes over the cohesive zone which makes the solution very expensive (elements of about 300 mm minimum side length and time steps of a mean value of 4 days were used).

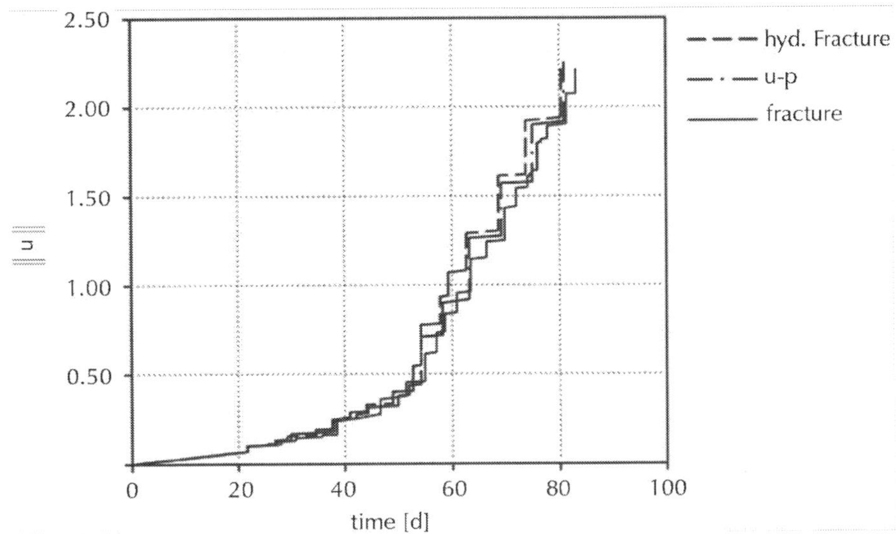

Figure 15: Relative displacements versus time at the dam crest.

In all three examples of hydraulic fracturing, the fracture site is not easily accessible. However, the fact that the effects of the stepwise advancement can be felt also at distance as shown in two examples would make them possible to be monitored remotely. Field data and experimental evidence on reservoir rocks and large bodies are still missing. Data could possibly come from fracking sites or from some fracking-induced earthquakes [3].

CONCLUSIONS

A fully coupled model for pressure-induced cohesive fracture in a saturated porous medium and its solution by the finite element method has been shown. The model is of the discrete crack type and requires continuous updating of the mesh as the crack tip advances. This is achieved with powerful mesh generators. Three types of refinement are necessary to obtain mesh-independent results: a refinement over the domain of the Zienkiewicz-Zhu type, an element threshold number over the process zone and a refinement in time, here with DGT. The results show that in case of pressure induced fracture with pressure exchange and flow between the fracture and the surrounding medium the crack tip advances stepwise. This was found also by few other authors. Smooth diagrams are found on the contrary in a thermo-elastic fracture which is a coupled problem but with stress-free fracture surfaces. From a comparison of results obtained with different methods by other authors, it appears that in some situations a particular adopted method hides the problems discussed in this paper because the required refinements clash with the *raison d'être* of such methods like XFEM or PUFEM (adoption of rough meshes). This is the reason why 'the physical phenomenon challenges the numerical scheme' [18] and why several authors dealing with hydraulic fracturing have not noticed the peculiar behaviour shown here. Also, two-step procedures may introduce some bias in the solution. Two different explanations are found in the literature for the discussed phenomena: one invokes pressure drop [2] and the other pressure peak [24] after crack advancement. The respective loading conditions are different, but the question deserves further scrutiny. Finally, the stepwise advancement may be relevant for earthquake engineering, see e.g. the resemblance between the obtained data of [2] and magnitude records of earthquakes. In many earthquake-prone regions, there is plenty of water available at the level where the rupture takes place [38,39]. The problem solved in [17] has been solved again with XFEM and finer mesh in [40] and the steps in the fracture advancement featured in [17] disappeared. This implies that XFEM yields a smooth solution for a phenomenon which in nature is not smooth: as shown in [2] hydraulic fracturing exhibits avalanche behaviour and hints of Self-Organized Criticality.

Appendix

The flow chart of Figure 16 shows the numerical procedure adopted with particular emphasis on the refinement part.

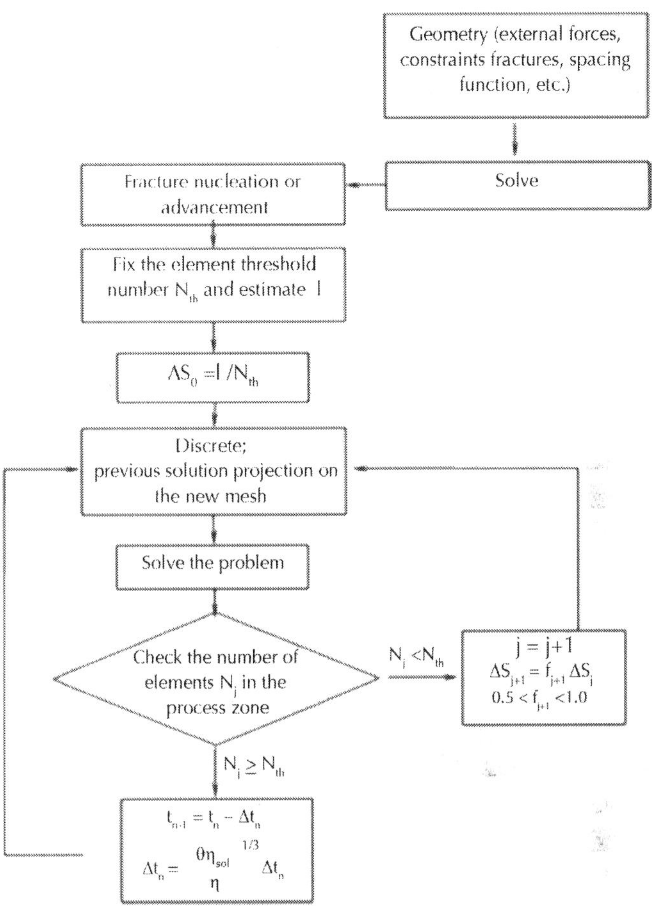

Figure 16: Flow chart.

AUTHORS' CONTRIBUTION

SS developed the code, devised the crack tip advancement procedure and carried out the simulations BS drafted the manuscript and explained the results in light of new findings in literature. All authors read and approved the final manuscript.

REFERENCES

1. Vidic RD, Brantley SE, Vandenbossche JM, Joxtheimer D, Abad JD (2013) Impact of shale gas development on regional water quality. Science 340:826-836
2. Tzschichholz F, Herrmann HJ (1995) Simulations of pressure fluctuations and acoustic emission in hydraulic fracturing. Phys Rev E 5:1961-1970
3. Ellsworth WL (2013) Injection induced earthquakes. Science 341:142-150
4. Perkins TK, Kern LR (1961) Widths of hydraulic fractures. SPE J 222:937-949
5. Rice JR, Cleary MP (1976) Some basic stress diffusion solutions for fluid saturated elastic porous media with compressible constituents. Rev Geophs Space Phys 14:227-241
6. Cleary MP (1978) Moving singularities in elasto-diffusive solids with applications to fracture propagation. Int J Solids Struct 14:81-97
7. Huang NC, Russel SG (1985) Hydraulic fracturing of a saturated porous medium—I: general theory. Theor Appl Fract Mech 4:201-213
8. Huang NC, Russel SG (1985) Hydraulic fracturing of a saturated porous medium—II: special cases. Theor Appl Fract Mech 4:215-222
9. Detournay E, Cheng AH (1991) Plane strain analysis of a stationary hydraulic fracture in a poroelastic medium. Int J Solids Struct 27:1645-1662
10. Advani SH, Lee TS, Dean RH, Pak CK, Avasthi JM (1997) Consequences of fluid lag in three-dimensional hydraulic fracture. Int J Num Anal Methods Geomech 21:229-240
11. Garagash D, Detournay E (2000) The tip region of a fluid-driven fracture in an elastic medium. J Appl Mech 67:183-192
12. Boone TJ, Ingraffea AR (1990) A numerical procedure for

simulation of hydraulically driven fracture propagation in poroelastic media. Int J Num Ana Methods Geomech 14:27-47

13. Carter BJ, Desroches J, Ingraffea AR, Wawrzynek PA (2000) Simulating fully 3-D hydraulic fracturing. In: Zaman M, Booker JR, Gioda G (eds) Modeling in geomechanics, Wiley, Chichester. pp 525-567
14. Schrefler BA, Secchi S, Simoni L (2006) On adaptive refinement techniques in multifield problems including cohesive fracture. Comp Methods Appl Mech Engrg 195:444-461
15. Secchi S, Schrefler BA (2012) A method for 3-D hydraulic fracturing simulation. Int J Fracture 178:245-258
16. Réthoré J, de Borst R, Abellan MA (2008) A two-scale model for fluid flow in an unsaturated porous medium with cohesive cracks. Comput Mech 42:227-238
17. Mohammadnejad T, Khoei AR (2013) Hydromechanical modelling of cohesive crack propagation in multiphase porous media using extended finite element method. Int J Numer Anal Meth Geomech 37:1247-1279
18. Kraaijeveldt F, Huyghe JM, Remmers JJC, de Borst R (2013) 2-D mode one crack propagation in saturated ionized porous media using partition of unity finite elements. J Appl Mech 80:020907-1-12
19. Secchi S, Simoni L, Schrefler BA (2004) Cohesive fracture growth in a thermoelastic bi-material medium. Comput Struct 82:1875-1887
20. Kraaijeveld F (2009) Propagating discontinuities in ionized porous media, Dissertation. Eindhoven University of Technology.
21. Kraaijeveldt F, Huyghe JM, Remmers JJC, de Borst R, Baaijens FPT (2014) Shearing in osmoelastic fully saturated media: a mesh-independent model. Engineering Fracture Mechanics. in press
22. Barenblatt GI (1959) The formation of equilibrium cracks during brittle fracture: general ideas and hypotheses: axially-

symmetric cracks. J Appl Math Mech 23:622-636

23. Remij EW, Pizzoccolo F, Remmers JJ, Smeulders D, Huyghe JM (2013) Nucleation and mixed-mode crack propagation in porous material. ASCE, Poromechanics V. pp 2260-2269 doi:10.1061/9780784412992.247
24. Pizzocolo F, Hyughe JM, Ito K (2013) Mode I crack propagation in hydrogels is stepwise. Eng Fract Mech 97:72-79
25. Dugdale DS (1960) Yielding of steel sheets containing slits. J Mech Phys Solids 8:100-104
26. Hilleborg A, Modeer M, Petersson PE (1976) Analysis of crack formation and crack growth in concrete by means of fracture mechanics and finite elements. Cem Concr Res 6:773-782
27. Camacho GT, Ortiz M (1996) Computational modelling of impact damage in brittle materials. Int J Solids Struct 33:2899-2938
28. Lewis RW, Schrefler BA (1998) The finite element method in the static and dynamic deformation and consolidation of porous media. Wiley, Chichester.
29. Witherspoon PA, Wang JSY, Iwai KJE, Gale JE (1980) Validity of cubic law for fluid flow in a deformable rock fracture. Water Resour Res 16:1016-1024
30. Li XD, Wiberg NE (1998) Implementation and adaptivity of a space-time finite element method for structural dynamics. Comp Methods Appl Mech Engrg 156:211-229
31. Secchi S, Simoni L, Schrefler BA (2008) Numerical difficulties and computational procedures for thermo-hydro-mechanical coupled problems of saturated porous media. Comput Mech 43:179-189
32. Secchi S, Simoni L (2003) An improved procedure for 2-D unstructured Delaunay mesh generation. Adv Eng Softw 34:217-234
33. Secchi S, Simoni L, Schrefler BA (2007) Numerical procedure for discrete fracture propagation in porous materials. Int J Num Anal Methods Geomech 31:331-345

34. Zhu JZ, Zienkiewic OC (1988) Adaptive techniques in the finite element method. Com Appl Num Methods 4:197-204
35. Rank E, Katz C, Werner H (1983) On the importance of the discrete maximum principle in transient analysis using finite element methods. Int J Num Methods Engng 19:1771-1782
36. Bae JS, Krishnaswamy S (2001) Subinterfacial cracks in bimaterial systems subjected to mechanical and thermal loading. Eng Fract Mech 68:1081-1094
37. ICOLD (1999) Fifth international benchmark workshop on numerical analysis of dams. Theme A2, Denver, Colorado.
38. Doglioni C, Barba S, Carminati E, Riguzzi F (2013) Fault on-off fluids response. Geoscience Frontiers. pp 1-14 doi.org/10.1016/j.gsf.2013.08.004
39. Kelbert A, Schultz A, Egbert G (2009) Global electromagnetic induction constraints on transition-zone water content variations. Nature 460:1003-1006 doi:10.1038/nature08257
40. Mohammadnejad T, Khoei AR (2013) An extended finite element method for hydraulic fracture propagation in deformable porous media with the cohesive crack model. Finite Elements in Analysis and Design 73:77-95

Chapter 2

A Systematic Review of Enhanced (or Engineered) Geothermal Systems: Past, Present and Future

Katrin Breede, Khatia Dzebisashvili, Xiaolei Liu, and Gioia Falcone

Institute of Petroleum Engineering, Clausthal University of Technology, Agricolastraße 10, Clausthal-Zellerfeld 38678, Germany

ABSTRACT

Enhanced (or engineered) geothermal systems (EGS) have evolved from the hot dry rock concept, implemented for the first time at Fenton Hill in 1977. This paper systematically reviews all of the EGS projects worldwide, based on the information available in the public domain. The projects are classified by country, reservoir type, depth, reservoir temperature, stimulation methods, associated seismicity, plant capacity and current status. Thirty five years on from the first EGS implementation, the geothermal community can benefit from the lessons learnt and take a more objective approach to the pros and cons of 'conventional' EGS systems.

REVIEW

The currently used term 'enhanced or engineered geothermal system' (EGS) has its roots in the early 1970s when a team from Los Alamos National Laboratories began the hot dry rock (HDR) project at Fenton Hill (Cummings and Morris 1979; Tester et al. 1989; Brown 1997; Duchane1998). The concept is described in Potter et al. (1974). HDR was also known as hot fractured rock because of either the need to fracture the virtually impermeable formations or the presence of natural fractures in the hot reservoir (Wyborn et al. 2005; Goldstein et al. 2011) or as hot wet rock (HWR) when it was established that the formations were not completely dry but contained some fluids. The European EGS project at Soultz-sous-Forêts in France is an example of a HWR reservoir (Duchane 1998). Further nomenclature encountered in the literature include stimulated geothermal system, deep heat mining (Häring and Hopkirk 2002; Häring 2007) and deep earth geothermal. All of the above usually imply the use of petrothermal systems (Ilyasov et al. 2010; Gebo NDS2012a).

Schulte et al. (2010) defined the typical geological settings for EGS, varying from igneous (e.g. Iceland), metamorphic (e.g. Lardarello, Italy), magmatic (e.g. Soultz, France) and sedimentary (e.g. Groß Schönebeck and Horstberg, Germany).

According to Potter et al. (1974), the most suitable rock type for HDR is granite or other crystalline basement rock; temperatures should vary from 150°C to 500°C at depths in the order of 5 to 6 km, with an average flow rate over a 10-year reservoir lifetime of 265 l/s, with hydraulic fracturing achieving a contact surface area of approximately 16 km^2, an average thermal capacity of 250 MW_{th} that could be obtained from the surface heat exchanger, and with pressurized water entering at 280°C and leaving at 65°C. Based on these criteria, the potential electrical power that could be generated might amount to 50 MW_e at a net efficiency of 20%. Over the years, different definitions of EGS have been proposed, covering a broad variety of rock types, depth, temperature, reservoir permeability and porosity, type of stimulation technique involved, etc. Below are four examples of recent EGS definitions in the public domain.

- The Massachusetts Institute of Technology (MIT) led an interdisciplinary panel which defined EGS as 'engineered reservoirs that have been created to extract economical amounts of heat from low permeability and/or porosity geothermal resources. For this assessment, this definition has been adapted to include all geothermal resources that are currently not in commercial production and require stimulation or enhancement. EGS would exclude high-grade hydrothermal but include conduction dominated, low permeability resources in sedimentary and basement formations, as well as geopressured, magma and low grade, unproductive hydrothermal resources. Co-produced hot water from oil and gas production is included as an unconventional EGS resource type that could be developed in the short term and possibly provide a first step to more classical EGS exploitation' (MIT et al. 2006a).
- The Australian Geothermal Reporting Code Committee considered EGS as 'a body of rock containing useful energy, the recoverability of which has been increased by artificial means such as fracturing' (AGRCC 2010).
- Williams et al. (2011) proposed that 'EGS comprise the portion of a geothermal resource for which a measureable increase in production over its natural state is or can be attained through mechanical, thermal, and/or chemical stimulation of the reservoir rock. In this definition, there are no restrictions on temperature, rock type or pre-existing geothermal exploitation'.
- The BMU (2011) defines enhanced geothermal systems as creating or enhancing a heat exchanger in deep and low permeable hot rocks using stimulation methods. Following BMU's definition, EGS embraces not only HDR but also deep

heat mining, hot wet rock, hot fractured rock, stimulated geothermal systems, and stimulated hydrothermal systems.

Clearly, the geothermal community lacks a universal definition of EGS, which may simply be taken as 'unconventional geothermal systems', diverging significantly from the initial HDR concept. This lack of clarity may constitute a potential obstacle to the implementation of tailored subsidy programmes.

In this study, the MIT definition is adopted with the only difference that geopressured and magmatic systems and also co-produced hot water from hydrocarbon wells are excluded. The reasons for this particular choice are that the MIT report was (and still is) regarded as a milestone report towards the development of EGS; also, from an engineering point of view, it is perhaps one of the most comprehensive definitions. On the other hand, it does not enter into the details of the different stimulation approaches and associated consequences for different EGS systems. Recently, for example, Jung (2013) has reconstructed the background to contemporary EGS: from the original HDR concept based on multi-zone hydraulic fracturing in competent crystalline formations, through that of open-hole massive injection in naturally fractured crystalline formations and finally to the proposed multi-zone massive injection (with the objective of generating multiple wing cracks) in naturally fractured crystalline formations. As this review does not aim at a project-by-project evaluation of the geomechanics that occur during EGS stimulation, the modified MIT definition is considered to be suitable for generating the database proposed in this study.

Geopressured and magma systems were left out from this review because they typically have been excluded from past EGS cataloguing attempts, such as those proposed by European Geothermal Energy Council (EGEC) (2012) and GtV (2013).

EGS Milestones

During the last four decades, there have been some key milestones towards the development of EGS for heat production and electricity generation. The information that follows is based on the report by Tenzer (2001), supplemented by additional information:

- 1970: Proposals for the first EGS worldwide in Fenton Hill, Los Alamos, USA.
- 1973: First EGS experiments in Fenton Hill.
- 1974 to 1977: Feasibility studies for EGS projects in Japan.
- 1975: Start of preparations for the first scientific EGS pilot plant in Bad Urach, Germany.
- Since 1977: EGS feasibility studies for shallow depths at Falkenberg, Germany, Camborne School of Mines, Cornwall in the UK and Le Mayet, France.
- 1977: EGS Bad Urach - drilling starts.
- 1980 to 1986: EGS Bad Urach - deepening of the borehole to 3,488 m at 147°C and hydraulic tests for single borehole system.
- 1984 to 1985: Start of EGS; Neustadt-Glewe, as a pilot project for low enthalpy energy; to date, this is the warmest accessed hot water reservoir in Northern Germany.
- 1986: Start of the German-French EGS project at Soultz-sous-Forêts, France, as a joint European research EGS pilot plant.
- 1986 to 1991: First EGS experiments in Hijori and other locations in Japan.
- 1987: EGS Soultz - began drilling the first borehole to 2,000 m at 140°C and started the investigation of the crystalline basement in the Rhine-Graben.
- 1989: EGS Soultz - UK joins the project; formation of an industrial consortium for organized planning and operation of an EGS project in Europe.
- 1990: EGS Soultz - drilling of a second 2,000-m deep borehole and deepening of the first borehole to 3,500 m depth (at 160°C); geothermal reservoir identification; the second borehole was used as seismic observation borehole.
- 1991 to 1996: EGS Bad Urach - deepening of the borehole to a depth of 4,445 m at a temperature of 172°C; also performed intense borehole measurement programme.

- 1994–1995: EGS Soultz - deepening of the second borehole to a depth of 3,876 m, followed by a production test which saw the first steam production in Middle Europe from crystalline rocks; using massive stimulation and circulation tests with seismic monitoring and development of the downhole heat exchanger, a thermal power of 8 MW was achieved.
- 1996: Start of deep heat mining project in Basel, Switzerland - a pilot project for EGS in a modern urban environment.
- 1996 to 1997: EGS Bad Urach - development of a downhole heat exchanger by massive hydraulic fracturing; the largest EGS created worldwide; long-term (4 months) hydraulic circulation test; a thermal power of 11 MW was achieved.
- 1998 to 2000: EGS Soultz - deepening of the second borehole to 5,060 m at 201°C; hydraulic stimulation and seismic monitoring.
- 2001: Start of EGS Groß-Schönebeck, Germany, which was the first in situ geothermal laboratory for developing techniques for the exploration and usage of geothermal energy.
- 2003: Start of EGS Cooper Basin, Australia - the largest demonstration EGS project in the world.
- 2003: Test of new single well concept in Genesys Horstberg, Germany.
- 2003: Start of EGS Landau - the first geothermal combined heat power plant to be connected to the grid; the one and only EGS project in a German town.
- 2004: Start of Unterhaching, Germany, the first geothermal project in the Bavarian Molasse Basin where, in addition to heat supply, electricity generation was also achieved; first Kalina power plant in Germany; first project worldwide with a private sector insurance for geological risk in deep boreholes.
- 2005: Start of EGS Paralana trying to implement an underground heat exchanger called 'heat exchanger within an insulator (HEWI)' concept (heat exchanger within the insulator) (Petratherm2012).
- 2006/2007: Deep heat mining project in Basel stopped due to repeated severe induced seismicity events; the project was

permanently abandoned in 2009.

- 2007: First binary geothermal plant in France at EGS Soultz (with ORC plant).
- 2009: New law for renewable energies in Germany - electricity generation and supply to the power net gets more financial support.
- 2009: Start of EGS GeneSys Hannover, Germany, as a single well concept.
- 2009: Start of EGS St. Gallen, Switzerland.
- 2010: Implementation of new 'side-leg' concept in the EGS project Insheim (Germany); forked injection well shall reduce induced seismicity (Insheim 2012).
- 2011: EGS GeneSys Hannover put on hold due to salt deposition in the single well.
- 2011: Guidelines for 'seismic surveillance' for Germany published by Bundesverband Geothermie.
- 2012: Switzerland decides to support deep geothermal projects.
- 2012: EGS Insheim connected to the power net.
- 2013: EGS Habanero successfully commissioned, with generation of 1 MW_e of power; first EGS project in Australia generating electricity.
- 2013: EGS St. Gallen drilling started.
- 2013: EGS St. Gallen put on hold due to induced seismicity events with a maximum magnitude of 3.6 on the Richter Scale; green light to proceed given by City Council 5½ weeks later.

Systematic Overview of Past and Present EGS Projects Worldwide

The following review should not be considered exhaustive as it is based exclusively on the information available in the public domain. Yet, to the authors' knowledge, this is the first public attempt to formally collate a large database of information on EGS worldwide, from the first HDR project at Fenton Hill in 1974 to date.

The objective of this review is to present key information on past and present EGS experience worldwide, from which key lessons can be learnt for the future.

The 31 EGS projects identified during this review are classified by country, reservoir type, depth, reservoir and wellhead temperature, stimulation methods, induced seismicity and radioactivity, plant capacity, flow rate and current status.

The 31 projects are divided into four different groups:

Table 1 comprises basic information about EGS projects that are still under development. It does not include pending commercial projects that are either at the status of raising funds (e.g. Munster in Germany and Eden in the UK) or still need governmental approval.

Table 1: EGS projects (R&D and commercial) still under development and not generating electricity

Project	Start date	Location	Well depth (m)	Stimulation methods	Description	Operator	Current status	Rock type	BHT (°C)	Seismic event	Flow rate (l/s)
Le Mayet[a]	1978 (Cornet2012)	France (Cornet 2012)	200 to 800 (Cornet 2012)	Hydraulic fracturing with and without proppant (Cornet2012; MIT et al.2006**b**)	Research (Cornet 2012; MIT et al. 2006**b**)	Not known	Not known	Granite (Cornet2012)	22 (Wyborn2011)	Microseismic, not felt on surface (Cornet2012)	5.2 (Wyborn2011)
Genesys Hannover	2009 (Zimmermann et al. 2009)	Germany (Zimmermann et al. 2009)	3,900 (Zimmermann et al. 2009)	Hydraulic fracturing (Zimmermann et al. 2009)	Demonstrate single well concepts (Zimmermann et al. 2009)	Federal Ministry of Economics and Technology (Zimmermann et al. 2009)	Salt deposition has been removed (BGR 2013)	Bunter sandstone (Zimmermann et al. 2009)	160 (Blöscher et al. 2012)	Microseismic (1.8 M) (Huenges2010)	7 (planned) (Zimmermann et al. 2009)
Groß Schönebeck	2000 (Zimmermann et al. 2009)	Germany (Zimmermann et al. 2009)	4,309 (Zimmermann et al. 2009; BINE 2012a) to 4,400 (BINE 2012**a**)	Hydraulic gel proppant and fracturing (Zimmermann et al. 2009; Blöscher et al. 2012; Huenges 2010) thermal (ENGINE2008b), chemical (Henninges et al.**2012**)	1st *in situ* geothermal laboratory, EGS research (Zimmermann et al. 2009)	GFZ, Schmidt + Clemens GmbH + Co. KG (BINE2012a)	Production-injection experiment and data interpretation and modelling finished (Feldbusch et al. 2013)	Sandstone and andesitic volcanic rocks (Zimmermann et al. 2009; Blöscher et al.2012)	145 (Blöscher et al. 2012)	Negligible (max, −1.8 to −1.0M) (Blöscher et al. 2012)	20 (Blöscher et al. 2012)

Mauerstetten	2011 (Schrage et al. 2012a)	Germany (Schrage et al. 2012a)	4,545 (Exorka2013)	Chemical (Schrage et al. 2012b); hydraulic (Informationsportal Tiefe Geothermie2013a)	Research (Schrage et al. 2012b)	Exorka GmbH, GFZ, TUBAF (Schrage et al. 2012a)	Seismic monitoring system installed (Informationsportal Tiefe Geothermie2013a); next step, hydraulic stimulation (Informationsportal Tiefe Geothermie2013a)	Limestone (Schrage et al.2012a)	130 (Schrage et al. 2012a)	Unknown	Unknown
St. Gallen	2009 (Geothermie Stadt St. Gallen 2013a)	Switzerland (Geothermie Stadt St. Gallen 2013a)	4,450 (Geothermie Stadt St. Gallen 2013a)	Chemical and hydraulic (Geothermie Stadt St. Gallen 2013a)	Commercial: heat and power (Geothermie Stadt St. Gallen 2013a)	ITAG Tiefbohr GmbH (Geothermie Stadt St. Gallen 2013a)	Production test interrupted due to pump failure and resulting seismic event (Geothermie Stadt St. Gallen2013a)	Malm, shell limestone (Geothermie Stadt St. Gallen2013a)	130 to 150 (estimated) (Geothermie Stadt St. Gallen2013a)	3.5 M (Geothermie Stadt St. Gallen2013a)	(Geothermie Stadt St. Gallen 2013a)
Newberry	2010 (Cladouhos et al. 2012)	USA (Cladouhos et al. 2012)	3,066 (BLM2012)	Hydroshearing, multi-zone isolation techniques (Cladouhos et al.2012)	Demonstration for EGS stimulation/research (Cladouhos et al. 2012)	AltaRock Energy, Davenport Newberry (Cladouhos et al. 2012)	Stimulation started successfully (Informationsportal Tiefe Geothermie2012)	Volcanic rocks (Fittermann1988)	315 (Cladouhos et al. 2012)	Micro-seismic (Cladouhos et al. 2012)	Unknown
Northwest Geysers	In 1980s (Garcia et al.2012)	USA (Romero et al. 1995)	3,396 (Garcia et al. 2012)	Thermal fracturing (Walters 2013)	Demonstration/research (Garcia et al. 2012)	Calpine Corporation (Garcia et al.2012)	Stimulation stage (5 MW of potential production) (Walters 2013)	Metasedimentary rocks (greywacke) (Romero et al.1995; Garcia et al. 2012)	About 400 (Garcia et al. 2012)	Microseismic (0.9 to 2.87 M) (Garcia et al. 2012; Walters2013)	9.70 (Garcia et al. 2012)

Paralana	2005 (Petratherm2012)	Australia (Petratherm2012)	4,003 (Petratherm2012)	Hydraulic (Petratherm 2012)	Commercial power development (Petratherm 2012)	Petratherm, Beach Energy (Petratherm2012)	Drilling of Paralana 3, submit funding application (Petratherm 2012)	Metasediments, granite (Petratherm2012)	171 (Petratherm2012)	Microseismic ≤2.6 M (Petratherm2012)	Up to 6 (ENGINE2008b)

GFZ, German Research Centre for Geosciences; TUBAF, Technische Universität und Bergakademie Freiberg (Germany); BHT, bottomhole temperature; [a]Note that little and contrasting information was found in the open domain concerning the project 'Le Mayet'. Some sources say that it was operational from 1984 till 1987 (Evans 2011). Others (MIT et al. 2006b) report that it was still ongoing as of 2006 and having a BHT of 33°C instead of the 22°C reported in the table.

Breede *et al.*

Breede *et al. Geothermal Energy* 2013 1:4, doi: 10.1186/2195-9706-1-4

Tables 2 and 3 present projects that are already in the power generation phase.

Table 2: Ongoing EGS projects (R&D and commercial) generating electricity

Project	Start date	Location	Well depth (m)	Stimulation methods	Description	Operator	Rock type	Reservoir temperature (°C)	Seismic event
A. Bruchsal	1,983 (BMU2011)	Germany (BMU2011)	1,874 to 2,542 (BMU2011)	Unknown	Commercial (Enbw 2013)	EnBW, EWB (KIT2013)	Bunter Sandstone (KIT 2013)	123 (Rettenmaier2012)[a];	Microseismic (KIT 2013)
Landau	2003 (BINE2012d)	Germany (Baumgärtner2012)	3,170 to 3,300 (Baumgärtner2012)	No stimulation for producer; hydraulic for injector (Baumgärtner2012)	First implementation of EGS technology in Germany (BINE2012d); first and only EGS in town in (D) (Baumgärtner2012)	BESTEC, Geox (Baumgärtner2012)	Granite (Lacirignola and Blanc 2012)	159 (Baumgärtner2012)a	Microseismic (≤2.7 M) (Baumgärtner2012), felt by residents
Insheim	2007 (Insheim2012)	Germany (Insheim 2012)	3,600 to 3,800 (LGB-rlp 2012)	Yes (Baumgärtner2012)	New concept, side-leg injection well (BINE2012b)	Pfalzwerke geofuture GmbH (Pfalzwerke-geofuture 2012; BINE 2012b)	Keuper, perm, bunter sandstone, granite (Baumgärtner2012)	165 (LGB-rlp2012)	M: 2.0 to 2.4 and microseismic (Groos et al.2012)
Neustadt-Glewe	1984 (BMU2011)	Germany (Bracke 2012)	2,320 (Bracke2012)	Unknown	Commercial, pilot plant for low enthalpy (BMU2011)	WEMAG AG, Stadt Neustadt-Glewe, Geothermie Neubrandenburg GmbH (BMU2011)	Sandstone (BMU2011)	99 (GtV 2013)a	Unknown
Unterhaching	2004 (BMU2011)	Germany (Bracke 2012)	3,350 to 3,580 (Bracke 2012)	Acidizing (BMU2011)	First Kalina power plant in Germany (BINE2012c)	Geothermie Unterhaching GmbH & Co. KG, Rödl & Partner GbR (BINE2012c)	Limestone (Dumas 2010)	123 (Bracke2012)a	Unknown

Soultz	1987 (MIT et al. 2006b)	France (Genter2012)	5,093 (MIT et al.2006d)	Hydraulic fracturing and acidizing (MIT et al. 2006d)	Research and demonstration (Genter 2012)	European cooperation project (MIT et al.2006d)	Granite (MIT et al.2006d)	165 (BMU2011)	Microseismic (M = −2 to 2.9) (Genter2012)
Bouillante	1963/1996 (Bertini et al.2006)	France (Guadeloupe) (Bertini et al.2006)	1,000 to 2,500 (Bertini et al.2006)	Thermal cracking (Bertini et al.2006)	Commercial (Bertini et al.2006)	Geothermie Bouillante, CFG-Services, BRGM, ORKUSTOFNUN, COFOR (Bertini et al. 2006)	Volcanic lavas and tuffs (Bertini et al.2006)	250 to 260 (Bertini et al.2006)	Microseismic (Sanjuan et al.2010)
Altheim	1989 (Pernecker1999)	Austria (Bloomquist 2012)	2,165 to 2,306 (Bayerisches Landesamt für Wasserwirtschaft 2011)	Acidizing (Pernecker1999), hydraulic stimulation (ENGINE2008b)	Commercial (Pernecker1999)	Municipality of Altheim, Terrawat (Pernecker 1999)	Limestone (Bayerisches Landesamt für Wasserwirtschaft 2011)	106 (Bloomquist 2012)	Unknown
Lardarello	1970 (Cappetti2006) (1904)	Italy (ENGINE2008b)	2,500 to 4,000 (Bertini et al.2006)	Hydraulic and thermal stimulation (ENGINE2008b)	Research and demonstration (Cappetti 2006) and commercial	ENEL Green Power (Lazzarotto and Sabatelli 2005)	Metamorphic rocks (ENGINE2008b)	300 to 350 (ENGINE2008a'a	≤3.0 M (Bromley 2012)
Coso	2002 (Häring2007)	USA (Häring2007)	2,430 to 2,956 (Julian et al.2009)	Hydraulic, thermal and chemical (Rose et al. 2004)	Research and development (Häring 2007)	Coso Operating Company (EGS Coso 2013)	Diorite, granodiorite, granite (Rose et al. 2004)	≥300 (EGS Coso 2013)	≤2.8 M (Julian et al. 2009)
Desert Peak	2002 (MIT2006c)	USA (MIT et al.2006c)	About 1,067 (Chabora et al.2012)	Shear, chemical, hydraulic (Davatzes et al. 2012)	Research and development (Davatzes et al.2012)	Ormat, GeothermEx (Val Pierce 2011)	Volcanic and metamorphic rocks (Chabora et al. 2012)	179 to 196 (Chabora et al.2012)	Microseismic: −0.03 to 1.7 (Chabora and Zemach 2013)
Berlín	2001 (Bommer et al. 2006)	El Salvador (Rodríguez2003)	2,000 to 2,380 (Rodríguez 2008)	Hydraulic fracturing and chemical (Rodríguez2003)	Developing EGS project in a geothermal field (Rodríguez2003)	Shell International (Rodríguez2003), LaGeo (Bommer et al.2006)	Volcanic rocks (Häring 2007)	183 (Bommer et al. 2006)	≤4.4 M (Bommer et al.2006)

Cooper Basin	2003 (Majer et al. 2007)	Australia (Majer et al.2007)	4,421 (Majer et al.2007)	Hydraulic (Majer et al.2007; Holl2012)	Largest demonstration project in the world (Stephens and Jiusto 2010)	Geodynamics Ltd. (Majer et al.2007; Geodynamics2013)	Granite (Majer et al. 2007)	242 to 278 (Geodynamics2013)	≤3.7M (Majer et al. 2007)
Hijiori	1985 (Sasaki1998)	Japan (Sasaki1998)	1,805 to 1,910 (Sasaki 1998)	Hydraulic fracturing (Sasaki 1998)	Developing EGS technologies (Sasaki 1998)	Japan's new energy (DiPippo2012a), NEDO (Sasaki 1998)	Granodiorite (Sasaki 1998)	190 (DiPippo2012a)	Microseismic (Sasaki 1998)

[a]Reservoir temperature not available, BHT is used instead.

Breede *et al.*

Breede *et al. Geothermal Energy* 2013 1:4, doi:10.1186/2195-9706-1-4

Table 3: Ongoing EGS projects (R&D and commercial) generating electricity

Project	Type of power plant	Flow rate (l/s)	Distance between producer and injector (km)	Installed electrical capacity (MWe)	Thermal capacity (MWth)	Flow assurance problem
B. Bruchsal	Kalina cycle (BMU2011)	28.5 (BMU2011)	1.4 (BMU2011)	0.55 (BMU2011)	5.5 (GtV 2013)	High salt contents (100 g/l); high CO_2 concentration (BMU2011)

Landau	Organic Rankine Cycle (BINE 2012d)	70 to 80 (Baumgärtner2012)	1.5 (Bracke 2012)	Up to 3.6 (Baumgärtner 2012)	2 to 5 (Baumgärtner2012)	Unknown
Insheim	Organic Rankine Cycle (Informationsportal Tiefe Geothermie 2013b)	65 to 85 (planned) (Pfalzwerke-geofuture2012)	Unknown	4.8 (Pfalzwerke-geofuture 2012)	6 to 10 (Pfalzwerke-geofuture 2012)	Unknown
Neustadt-Glewe	Organic Rankine Cycle (Bracke 2012)	35 (Bracke 2012)	1.5 (BMU2011)	0.21 (Bracke 2012)	4 (Bracke 2012)	High salt content, high gas concentration (Bracke2012)
Unterhaching	Kalina cycle (BMU 2011)	150 (Bracke 2012)	4.5 (Bracke 2012)	3.36 (Bracke 2012)	38 (Bracke 2012)	Unknown
Soultz	Organic Rankine Cycle (BMU 2011)	30 (BMU 2011)	0.6 (BMU 2011)	1.5 (BMU 2011)	Non-scheduled (Dumas2010)	Corrosion due to high salt contents (BMU2011)
Bouillante	Double and single flash (Bertini et al. 2006)	150 (Bertini et al. 2006)	0.5 (Bertini et al. 2006)	15 (Bertini et al. 2006)	Unknown	Unknown
Altheim	Organic Rankine cycle (Bayerisches Landesamt für Wasserwirtschaft 2011)	81.7 (Bayerisches Landesamt für Wasserwirtschaft 2011)	1.7 (Bayerisches Landesamt für Wasserwirtschaft 2011)	1.0 (Bloomquist 2012)	12.4 (Bayerisches Landesamt für Wasserwirtschaft 2011)	Clogging by a mixture consisting of stone material and bentonite (Pernecker1999)

Lardarello	Not known	100 (Cappetti 2006)	Variable: generally > 0.5 (ENGINE 2008a)	700 (ENGINE 2008b)	Not known	Highly corrosive, total loss of circulation when very high permeability fracture zones are encountered (ENGINE2008b)
Coso	Unknown	Unknown	1.4 (Julian et al. 2009)	240 (Karner 2005)	Unknown	Unknown
Desert Peak	Unknown	100 (Chabora and Zemach 2013)	Unknown	1.7 (additional) (Chabora and Zemach2013)	Unknown	Wellbore instability due to chemical stimulation (Chabora et al. 2012)
Berlín	Binary power plant (Prevost2004)	Unknown	Unknown	54 (Bommer et al.2006)	56 (Rodríguez 2000)	Unknown
Cooper Basin	Unknown	30 (Holl 2012)	Unknown	1 (Geodynamics 2013)	Unknown	Unknown
Hijiori	Binary Power Plant (DiPippo2012a)	17 (Sasaki 1998)	0.038 to 0.063 (DiPippo2012a)	0.13 (DiPippo 2012a)	8 (DiPippo 2012a)	High water losses (Johansson et al.1993), precipitation of anhydrite (DiPippo2012a)

Breede *et al.*

Breede *et al. Geothermal Energy* 2013 1:4, doi: 10.1186/2195-9706-1-4

Table 4 gives information about experimental projects that were developed to test single phase of an EGS project rather than the whole process to generate electricity.

Table 4: Concluded experimental EGS projects (without power generation)

Project	Descr-iption	Start date	Location	Rock type	Reservoir temperature (°C)	Well depth (m)	Stimulation methods	Seismic event	Fluid temperature (°C)	End date	Flow rate (l/s)
Falkenberg	Investigation of hydraulic fracturing at shallow depth (Tenzer 2001)	1977 (Tenzer2001)	Germany (Tenzer2001)	Granite (MIT et al. 2006e)	13.5 (Kappelmeyer and Jung1987)	500 (Tenzer2001)	Hydraulic fracturing (Tenzer2001)	Microseismic (MIT et al.2006e)	Unknown	1986 (Tenzer2001)	0.2 to 7 (Kappelmeyer and Jung1987) (test)
Genesys Horstberg	Testing of new single well concepts at an abandoned gas well (BGR 2012a)	2003 (BGR 2012b)	Germany (BGR 2012a)	Sedimentary (BGR 2012a)	150 (ENGINE2012)	3,800 (ENGINE2012)	Hydraulic fracturing (BGR 2012a)	No measured event (Kreuter2011)	115 (Tenzer2001)	2007 (estimation) (BGR 2012b)	10 to 20 (Tischner et al. 2010)
Fjällbacka	Experimental project (Portier et al. 2007)	1984 (Jupe et al.1992)	Sweden (Portier et al. 2007)	Granite (Portier et al. 2007)	16 (Wallroth et al. 1999)	70 to 500 (Jupe et al. 1992)	Hydraulic fracturing and acidizing (Portier et al.2007)	Microseismic (Wallroth et al. 1999)	Unknown	1995 (Wallroth et al. 1999)	0.9 to 1.8 (Wallroth et al.1999)

Rosema-no-wes	Experimental project (MIT2006f)	1977 (MIT 2006f)	UK (MIT2006f)	Granite (MIT 2006f)	79 to 100 (MIT 2006f)	2,000 to 2,600 (MIT2006f)	Hydraulic fracturing (MIT 2006f), viscous gel stimulation (Parker1999), placement of proppants in joints (Parker1999)	Max. magnitude, 3.1 (Bromley and Mongillo2008)	54.2 to 80 (Richards et al. 1992)	1992 (MIT2006f)	4 to 25 (MIT2006f)
Fenton Hill	First EGS in the world (MIT 2006g)	1974 (MIT 2006g)	USA (MIT2006g)	Crystal-line rock (Brown2009)	200 to 327 (MIT 2006g)	2,932 to 4,390 (MIT2006g)	Hydraulic fracturing (MIT 2006g)	Microseismic (Brown 1995)	180 to 192 (MIT 2006g)	1993 (MIT2006g)	10.6 to 18.5 (MIT 2006g)
Ogachi	Test run EGS project in shallow depth (Kaieda et al.2005)	1989 (Kaieda et al.2005)	Japan (Kaieda et al. 2005)	Granodiorite (Kaieda et al.2010)	60 to 228 (Kaieda et al.2005)	400 to 1100 (Kaieda et al.2005)	Multiple wells with mul-tiple fracture zones (Kaieda et al.2005); hydraulic (Kaieda et al.2005)	Few microseismic (Kaieda et al.2010)	160 (test result) (Kaieda et al. 2005)	2002 (Kaieda et al. 2005)	6.7 to 20 (Kaieda et al.2005) (test)

Breede *et al.*

Breede *et al.* *Geothermal Energy* 2013 1:4, doi:10.1186/2195-9706-1-4

Table 5 presents information on projects that are aimed for electricity generation but were abandoned due to various problems. Input information was drawn from different sources available in the public domain; all of which are cited in the titles of the tables.

Table 5: Abandoned or on hold EGS projects

Project	Operator	Description	Start date	Location	Rock type	Stimulation methods	Seismic event	Well depth (m)	End date	Reasons of abandonment
Bad Urach	Forschungs-Kollegium Physik des Erdkörpers (MIT 2006h)	EGS pilot by one borehole only (Tenzer 2001)	1977 (Tenzer2001), 2006 (Wyborn2011)	Germany (Tenzer 2001)	Gneiss (Tenzer et al.2000)	Hydraulic fracturing (Schanz et al.2003)	Micro-seismicity (Schanz et al.2003)	3,334 to 4,445 (Schanz et al. 2003)	1981 (MIT2006i), 2008 (Wyborn2011)	Torn off bore rods in borehole (Wyborn 2011)
Basel	Geopower Basel (Romano2009)	Planning to develop EGS project (Ladner and Häring2009)	1996 (Giardini2009)	Switzerland (Romano2009)	Granite (Ladner and Häring 2009)	Hydraulic fracturing (Ladner and Häring 2009)	Frequent earthquakes (including 3.4 M) (Romano 2009)	5,000 (Romano2009)	2009 (Giardini2009)	Induced seismicity exceeding acceptable levels (Giardini 2009)
The Southeast Geysers	AltaRock Energy (Romano2009)	Redrill a well for EGS demonstration project (AltaRock Energy Inc. 2012)	2008 (Cotler2009)	USA (Romano2009)	Greywacke (AltaRock Energy Inc.2012)	Multiple fractures zones in wells (planned) (AltaRock Energy Inc.2012)	Induced seismicity risk (Romano 2009)	1,341 (AltaRock Energy Inc.2012)	2009 (AltaRock Energy Inc.2012)	Wellbore collapsing and induced seismicity risk (Romano2009)

Breede *et al.*

Breede *et al.* *Geothermal Energy* 2013 1:4, doi:10.1186/2195-9706-1-4

This grouping criteria allow the reader to have an immediate overview of past vs. current vs. future EGS activities, better appreciate the challenges faced by EGS (technical, economic and related to public acceptance), develop a feeling for the level of research and development efforts put into EGS vis-à-vis the desire to achieve worldwide commercialisation of the concept.

Note that in the tables, 'microseismic' refers to seismic activity less than 3.5 on the Richter scale and is used for those cases when no further details on recorded seismicity could be found in the literature. See the following paragraphs for more discussions on induced seismicity in EGS projects.

When 'thermal capacity' is quoted next to 'installed electrical capacity', this implies a combined heat and power project.

Under 'stimulation methods', the terms 'hydraulic fracturing', 'hydraulic', 'hydroshearing', 'shear' and 'hydraulic stimulation' are taken directly as quoted by the cited sources. The authors of this manuscript have not performed an independent review or assessment of the specific stimulation methods implemented in or planned for each individual project, as this falls outwith the scope of this broader EGS review.

Overall, the tables above capture a detailed database of 31 EGS projects worldwide. Based on the tables, the following plots provide a way to extract trends and common characteristics of EGS.

As illustrated in Figure 1, most of the European EGS projects' reservoir/bottomhole temperatures are lower than 165°C, with the exception of Lardarello and Bouillante. Compared to Europe, the average EGS reservoir/bottomhole temperatures in America, Australia and Asia are higher although the well depths are comparable. Note that only 25 projects are displayed in Figure 1; the remaining 6 projects (St. Gallen, Fjällbacka, Falkenberg, The Southeast Geysers, Basel and Bad Urach) are excluded because reservoir/bottomhole temperature data could not be found in the public domain or are only estimated in the case of St. Gallen.

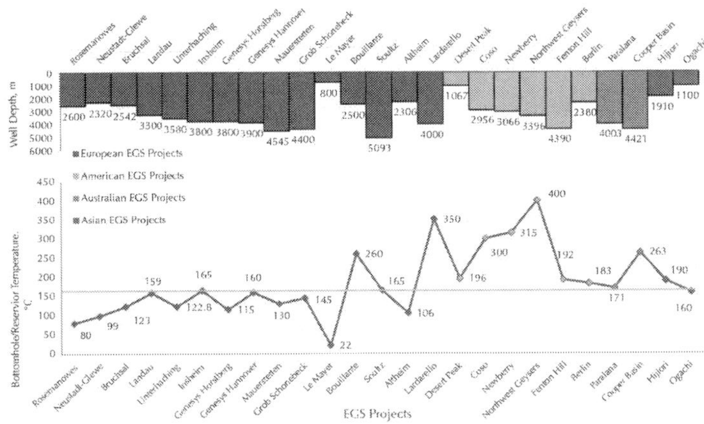

Figure 1: Worldwide EGS projects' reservoir/bottomhole temperature vs. depth.

The relationship shown in Figure 2 points out that most EGS activities are operated at flow rates lower than 40 l/s. Note that only 20 projects are displayed in Figure 2; the remaining 11 projects (Genesys Hannover, Insheim, Mauerstetten, Newberry, Coso, Berlín, Falkenberg, The Southeast Geysers, Basel, Bad Urach and St. Gallen) are excluded because flow rate data could not be found in the public domain.

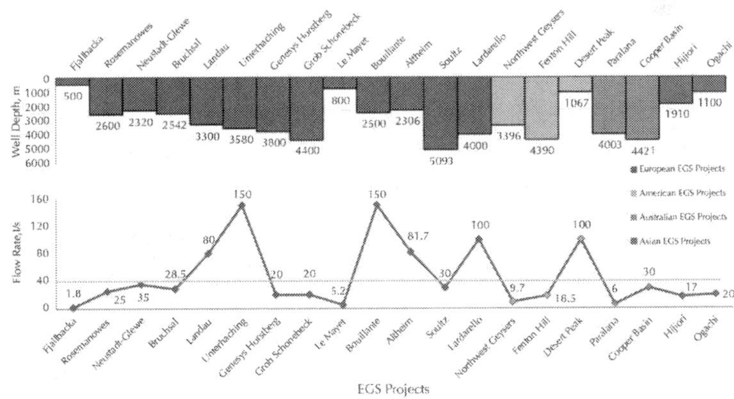

Figure 2: Worldwide EGS projects' flow rate vs. depth.

Figure 3 displays EGS projects classified on the basis of rock types. Although it appears that EGS activities can be implemented in any of the three major groups of rocks on earth, most projects are developed in igneous rocks, following the original HDR concept.

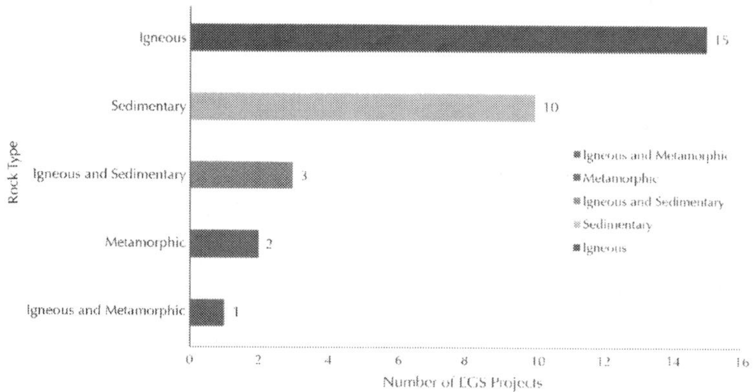

Figure 3: EGS projects classified on the basis of rock types.

The recorded maximum magnitudes of induced seismic events associated with the development of EGS projects worldwide are shown in Figure 4. Originally, the Richter scale was developed as a mathematical device to compare local earthquake sizes. The magnitude is defined as the logarithm of the wave amplitude recorded by seismographs. At that time, the smallest measurable earthquakes were assigned with values close to zero. However, due to the higher accuracy of modern seismographs, the Richter scale now measures earthquakes having negative magnitudes. Majer et al. (2007) reported that '...To date, the maximum observed earthquakes attributed to EGS operations have been magnitude 3.0 to 3.7 and the largest geothermal injection-related event was magnitude 4.6'. Later, Majer et al. (2013) also stated that for EGS, earthquakes are typically smaller than M 3.5 (M representing the momentum magnitude in this context). According to EGEC (2013), microseismic activity is less than 3.5 on the Richter scale. Only the projects with published induced seismic magnitude are displayed in Figure 4; the remaining 16 projects (Le Mayet, Mauerstetten,

Newberry, Bruchsal, Neustadt-Glewe, Unterhaching, Bouillante, Altheim,, Hijiori, Genesys Horstberg, Fjällbacka, Fenton Hill, Ogachi, Bad Urach, Falkenberg and The Southeast Geysers), most of which have been reported to suffer from microseismicity, are omitted due to lack of explicit seismic data.

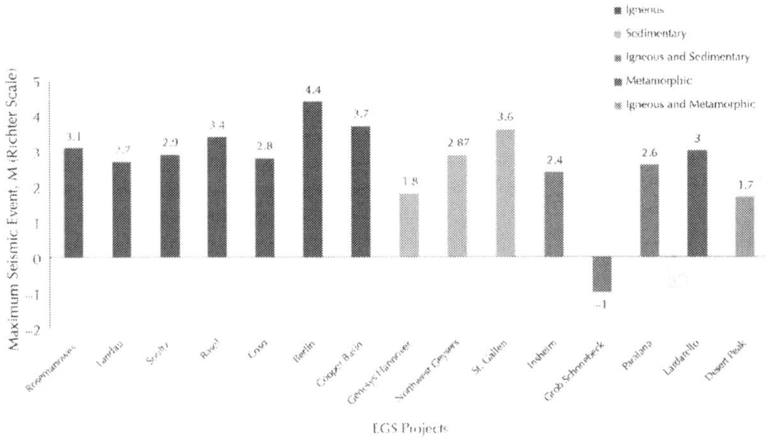

Figure 4: Projects with published induced seismic magnitude.

Stimulation methods that are applied in EGS developments are summarized in Figure 5, which reveals that hydraulic stimulation is the most commonly used method, independently of the rock type concerned. In addition, there are relatively few cases where chemical or thermal stimulation technologies are applied. This often leads to the assumption that the EGS definition only applies to hydraulically fractured systems.

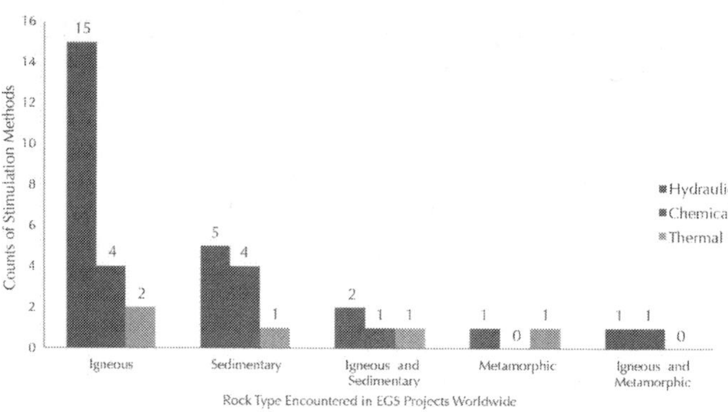

Figure 5: Stimulation methods applied to EGS projects worldwide.

The installed electrical and thermal capacity of EGS projects are summarized in Figure 6. Since EGS is still a developing concept, the database contains only 14 projects carried out with electricity generation. Note that the thermal capacities of Bouillante, Soultz, Lardarello, Desert Peak, Cooper Basin and Coso are missing as data could not be found in the public domain. The variation of production scale causes great capacity differences among the projects.

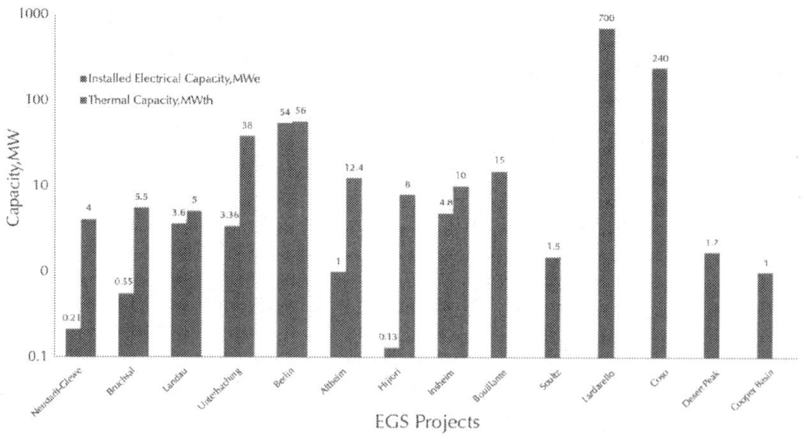

Figure 6: Installed electrical and thermal capacity of worldwide EGS projects.

Figure 7 shows the rock type and well depth of all the studied EGS projects worldwide.

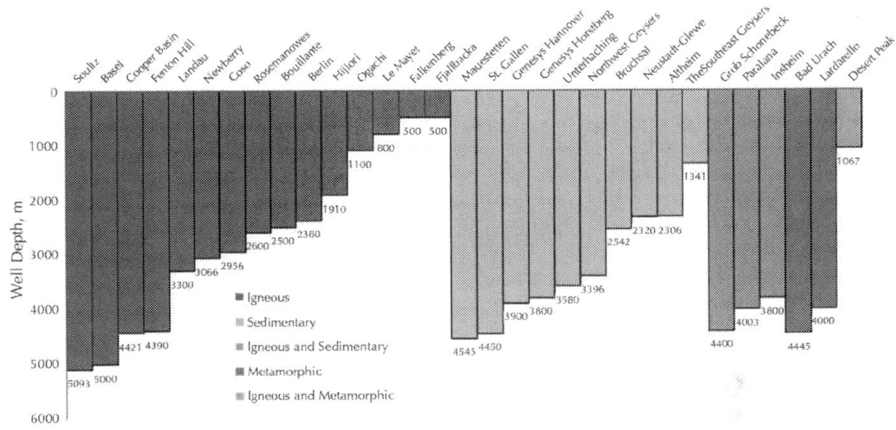

Figure 7: Rock type and well depth of EGS projects worldwide.

RESULTS AND DISCUSSIONS

From the information provided in the tables and the plots shown earlier, it appears that EGS projects currently under development are still on the learning curve, overcoming problems, gaining experience and trying to introduce advanced technology; the projects already concluded provide relevant history and analogy for upcoming developments and the projects that have been temporarily halted or abandoned give an insight into issues that must be avoided in the future.

Below are field cases where breakthrough methodologies were first implemented to validate the EGS concept. Unplanned events and issues that needed addressing in order to ensure feasibility and commerciality of EGS are discussed, and the corresponding lessons learnt are highlighted.

- The 'Paralana' project will use a new concept called the HEWI (Petratherm 2012).

- The 'Genesys' project was the first project worldwide testing a single well concept. Technical feasibility of the concept was proved by the 'Genesys Horstberg' project. The subsequent 'Genesys Hannover' aimed to use geothermal energy to heat the building complex of the Geozentrum Hannover (Tischner et al. 2010). The project has currently solved the problem of salt deposition, which has led to a suspension of the production test (Genesys 2012). This single well concept has the advantage of lower drilling costs as only one wellbore is needed to be drilled. However, since the circulating fluid moves through fractures, it is in direct contact with the rock formation, which leads to salt deposition risk. This experience has taught the geothermal community that flow assurance needs to be addressed ahead of time to prevent issues triggered by the chemical interaction between the injected fluid and the receiving rock, which can impair the overall success of an EGS project.
- The 'Groß Schönebeck' project is an important pilot for the development of geothermal technologies in Europe as an in situ laboratory was installed in one of the boreholes. The reservoir can be investigated by logging tools during production using a special Y-tool which is attached to the production string (Henninges et al. 2012). This system allows measurements with electrical tools and fibre-optic distributed temperature sensing. However, the data transfer to surface is problematic and can only be done discontinuously (Huenges 2013). This suggests that further technology advances are needed in the area of well logging for this type of applications. Using a newly developed fluid monitoring system, fluid physicochemical properties were measured online and in situ (Feldbusch et al. 2013). Worldwide, it is the only facility for the investigation of sedimentary large-scale structures under natural conditions. A 7-day long-term production-injection experiment between both boreholes to investigate the sustainability of the reservoir was completed in April 2012 (Feldbusch et al. 2013), and results, data interpretation and modelling of the experiments

have been presented by Feldbusch et al. (2013), Cherubini et al. (2013) and Noack et al. (2013). A corrosion test will permit the verification of the long-term reliability of the system's components Bine 4 (2012). A thermal fluid loop as the initial phase of fluid production was established and continuously operated for 7 days (Feldbusch et al. 2013).

- The Altheim project in Austria uses a special working fluid, which was never used before - a non-flammable, non-corrosive fluid with no ozone depletion activity (Bloomquist 2012).
- The 'Fenton Hill' project was the first attempt to extract geothermal energy form hot dry rocks with low permeability in the history of EGS (MIT et al. 2006g). One of the main lessons learnt from the Fenton Hill project is that an engineered hot reservoir should first be created from the preliminary borehole and then by connecting the enhanced reservoir and the injection borehole with the production boreholes (Brown 2009).
- The 'Rosemanowes Quarry' project in the UK stemmed directly from the positive results from Fenton Hill. One of the most significant lessons learnt from this project is that natural fractures and engineered fractures are almost unrelated. The natural fracture network plays a more important role compared with hydraulically enhanced fractures (MIT et al. 2006f). Also, as reported by Jung (2013), until then, the basement had been regarded as a competent rock mass, realizing that in reality, the basement contains open natural fractures even at great depth led to the abandonment of the HDR multi-fracture concept and the adoption of the hydraulic stimulation EGS concept.
- The 'Fjällbacka' project in Sweden gives similar conclusions to the Rosemanowes project, i.e. that naturally fractured systems dictate the results of reservoir stimulation (Wallroth et al. 1999).
- The 'Falkenberg' project began in 1976 and was planned as a test site for HDR at shallow depths to better understand

the mechanical and hydraulic properties of fractures (Kappelmeyer and Jung1987). A power generation phase was never intended.

- The 'Ogachi' project in Japan, a five-spot well pattern (four producers, one injector), was planned to be used for geothermal energy extraction from a shallow depth reservoir, but due to financial problems, the multiple production well system was not tried out. However, several basic technologies were successfully developed for general EGS activities through the project, which were later applied in another EGS programme in the Cooper Basin, South Australia in 2002 (Kaieda et al. 2005).
- The 'Basel' project in Switzerland saw induced seismic events - some exceeding 3.0 in magnitude - which led to its suspension (Ladner and Häring 2009). The Basel area has a history of natural seismic activity; the city was severely damaged by a 6.7 magnitude earthquake in 1356, the largest seismic event ever recorded in Central Europe (Giardini 2009). However, following a 3-year study after the seismic events recorded in connection with the geothermal project activities, the Basel project was cancelled. Induced seismicity associated with water injection and particularly hydraulic fracturing activities (due to changing stress patterns in reservoir rocks) has caused wide concern among the public (Majer et al. 2011).
- The 'Insheim' project has also had issues of induced seismicity. A so-called side-leg concept for the injection well was implemented to solve the problem (BINE 2012b). This concept enables pressure distribution during fluid injection over two separated ends of the injection well, thus minimizing the risk of induced seismicity. However, in 2013, another induced seismic event with a magnitude of 2.0 on the Richter scale occurred due to a water circulation stop during a reparation phase of the defective production pump (Geothermie-Pfalz 2013).
- The 'Landau' project is the first EGS project in a town in Germany, which is facing similar problems to Basel. Seismic

events of 2.7 in magnitude took place in 2009, which resulted to the temporary suspension of the operations. The project was restarted after purchasing 50 million of annual liability insurance to cover potential seismic damages (DiPippo 2012b). As a consequence of these events, water has to be reinjected at a reduced pressure to avoid induced seismicity, resulting in reduced power generation. The problem is planned to be tackled by implementing in 2013 the same side-leg concept that was used in Insheim (BINE 2012b).

- The 'Soultz-sous-Fôrets' project in France has allowed significant experience to be gained by several countries who participated in this joint project. Many experiments were conducted during the first 21 years of the project's life before the power plant was built. Different stimulation techniques, such as hydraulic fracturing with and without proppants and chemical stimulation were applied. Chemical stimulation has resulted in less seismic activity than other methods. Change of hydraulic parameters due to fracturing has resulted in an instantaneous variation of seismic activity. Seismic events with magnitudes greater than 2 have occurred during the shut-in phase. Although minor damages were caused by this EGS project, it did generate concern among the local population.

Microseismic monitoring has become an indispensable technology for the acceptance of EGS developments as it is the case for other applications of hydraulic fracturing and high-pressure water circulation (e.g. the exploitation of unconventional oil and gas resources). The experience gained from preliminary projects has led to a common view that induced seismicity associated with EGS activities can halt further development of this concept particularly in densely populated areas. More recently, though, despite the 3.6 magnitude seismicity induced by well control operations during drilling in St. Gallen, the city council decided to continue with the project and complete the first drilling phase (Geothermie Stadt St. Gallen 2013b).

- The 'Bad Urach' project suspended operations because of financing problems arising from a 'difficult geologic situation' at the well site, which indicated that this project would be unprofitable (DiPippo 2012b).

- The 'Geysers' project was abandoned due to drilling difficulties and the risk of increasing seismic activity (AltaRock Energy Inc. 2012).

It is worth mentioning here that according to Gebo NDS (2012b), drilling expenditure is the highest component in the development costs of an EGS project and can vary from 42% to 90% of the overall capital costs.

As mentioned earlier, hydraulic stimulation is the most commonly used technique for improving the permeability of a geothermal reservoir. Some of the world's EGS projects can extract geothermal energy from naturally fractured reservoirs, such as Northwest Geysers, Landau, Insheim, Urach, Bruchsal, Soultz-sous-Fôrets, Fjällbacka, Hijiori, Rosemanowes, Falkenberg and Newberry. The pre-existing naturally fractured networks can be stimulated by low pressure that is just above the critical pressure of shear failure (hydraulic shearing). However, the process of hydraulic fracturing, which uses injected water at high pressure to crack the rocks, is also frequently used especially in granite. Compared with hydraulic fracturing with high injection pressure, hydraulic shearing can easily crack rocks with low pressure and keep the fractures open without requiring a propping agent. Chemical stimulation, which is most applicable in carbonate rocks or used to dissolve carbonate cement in sandstone formations, along with thermal stimulation has also proved to be effective in some cases. However, there is relatively little literature concerning the application of chemical and thermal stimulation technologies in EGS projects.

Other issues associated with EGS stimulation are related to the potentially harmful effects on the surrounding environment. There has been public concern for the components of fracturing fluids that could represent a threat to drinking water sources. However, operators argue that EGS projects rarely require additives and chemicals (e.g. tracers, diverters, proppants) in fracturing fluids. When additives are necessary, then non-toxic chemicals are first considered. Also, deeply buried EGS reservoirs usually do not have a connection to near-surface groundwater aquifers, which would reduce the likelihood of contaminating drinking water (Regenspurg and Blöcher 2012).

Radioactivity is another problem emerging from EGS activities, which is caused by interaction between the geothermal fluid and certain formations containing radioactive elements. In general, the content of radionuclides in acidic magmatic rocks is higher compared to that in sedimentary rocks. Uranium and thorium are the most common radioactive elements found in granites. High reservoir temperature in EGS projects increases the solubility of radionuclides, which results in higher concentrations of these nuclides in the geothermal fluid. When the fluid is produced, the corresponding temperature reduction and pressure decrease in the surface facilities and causes deposition of scale, which leads to health, safety and environment problems. However, compared to other conventional energy production (e.g. oil and gas industry), the radioactivity occurring in EGS is likely to be very small (Battye and Ashman 2009). Radiation exposure of workers during the scale removal is avoided by using appropriate personal protection equipment. In general, the radiation exposure to the public is limited because long-lived natural radionuclides are not released during the operation of a geothermal power plant when the geothermal fluid is re-injected into the reservoir (Feige and Roloff 2012).

Almost all running EGS projects in the power generation phase utilize binary power plants. Binary systems use geothermal fluids with low temperature in the primary loop to vaporize working fluids with low boiling point that are used in the secondary loop to activate turbine-generator machine.

At present, two types of binary systems exist in the market: the organic Rankine cycle (ORC) which uses organic working fluids (e.g. propane or isobutane) and the Kalina Cycle which uses a mixture of two substances as the working fluid (e.g. water and ammonia). The advantage of the Kalina cycle over the ORC is that the abovementioned mixture boils at variable temperatures, which in turn creates higher efficiency at a certain inlet temperature, unlike the pure chemicals that are used in ORC (Clauser 2006).

The disadvantages of the Kalina system are the challenge of fine tuning the plant operation and the tendency of the ammonia-water mixtures to prematurely condense during expansion. Hence, the majority of the EGS projects implemented so far tend to use ORC power plants.

The flow rate recovered with EGS projects is crucial in dictating the success of a project. It needs to be high to ensure the project's economic viability. Yet if the rate is too high, there may not be sufficient 'residence time' for the circulating medium in the reservoir to extract enough heat from the rock. Depending on reservoir permeability, fracture surface area, pumping pressure, etc., the flow rate varies significantly from project to project.

Along with the ongoing debate over the definition of EGS, it has also been reported that the output of EGS projects is far lower than the theoretical expectation. Sanyal and Butler (2005) built a number of simulation models as a starting point for estimating EGS reserves on the basis of conditions seen at a desert park in USA, which suggested a recovery factor greater than 40% for EGS. However, Grant and Garg (2012) later pointed out that the recovery factor for the Cooper Basin EGS system would be lower than 2%, according to the modelled performance.

CONCLUSIONS

Many publications provide eye-catching numbers about EGS potential, yet there is still much to do to tap this energy. However, from this review of EGS projects worldwide, it transpires that EGS is still on a learning curve. Success is not guaranteed, and this implies significant financial risks for any EGS project, which can lead to its abandonment in some cases (e.g. Bad Urach project).

This observation leads to the natural question of why success is not guaranteed. From the classification exercise performed in this work, it is possible to conclude that the 'typical' EGS system does not exist, so much that, as shown in the introduction, the geothermal community does not even have a universally accepted and unambiguous definition of EGS as yet. The typical EGS system does not exist because - as shown in the tables and in the figures - there are several possible (and significantly different) geological, petrophysical, thermal, hydraulic and geomechanical environments where high temperature can be tapped underground. Even the depth where sufficiently high temperature can be encountered varies from region to region in the world, making it difficult to specify what 'deep geothermal energy' (another term often used within the geothermal community) really is and how it can be related to the EGS concept.

The problem is that of handling each particular EGS system in such a way that economic flow rates at the right temperature and over a sufficient time span can be obtained. It is commonly accepted that for an EGS doublet system to be of commercial size, assuming a depth greater than 3 km and a temperature greater than 150°C, the system should operate at flow rates between 50 and 100 l/s and produce an electric power of 3 to 10 MW_e over a life of at least 25 years (Jung 2013).

Based on the relatively limited EGS experience gathered to date and the extreme variety of natural occurrences and engineering solutions (including reservoir enhancement), it is therefore no surprise that EGS is still on a learning curve. This learning process must continue via more research and development, further technology advances and significantly more financial and political incentives before EGS will be commercially feasible, say in the next 10 to 20 years.

It is critical for EGS to ensure that relevant technologies are applied, having minimal risk of seismicity, and permitting the exploration of geothermal resource in a safe and environmentally friendly manner.

In the same vein, communities should be provided with regular, understandable and realistic information about EGS activities in order to gain public acceptance. Ongoing dialogue and interaction with communities are vital to achieve this.

AUTHORS' CONTRIBUTIONS

All: Systematic overview of past and present Egs projects worldwide + Results & Discussion. KB: Milestones + Tables. XL: Tables + Figures. KD: Review. GF: Conclusions. KB + GF: preparation of revised manuscript and rebuttal letter after referees' revision.

REFERENCES

1. AGRCC (Australian Geothermal Reporting Code Committee) (2010) Geothermal lexicon for resources and reserves definition and reporting. Australian Geothermal Reporting Code Committee, Adelaide.
2. AltaRock Energy Inc (2012) AltaRock EGS demonstration project status with NCPA at the Geysers. http://altarockenergy.com/AltaRock_EGS_Demonstration_Project_Status_101909.pdf . Accessed 19 Sept 2012
3. Battye DL, Ashman PJ (2009) Radiation associated with Hot Rock geothermal power. In: Budd AR, Gurgenci H (eds) Proceedings of the 2009 Australian Geothermal Energy Conference, Geoscience Australia, Record 2009/35, Commonwealth of Australia, Adelaide.17–19 November 2010
4. Baumgärtner J (2012) Insheim and Landau – recent experiences with EGS technology in the Upper Rhine Graben. Oral presentation presented at ICEGS 2012, Freiburg.25 May 2012
5. Bayerisches Landesamt für Wasserwirtschaft (2011) City of Altheim – Geothermal energy supply.http://ebookbrowse.com/27-slides-0-2-altheim-pdf-pdf-d 183736900 . Accessed 1 Nov 2012
6. Bertini G, Casini M, Gianelli G, Pandeli E (2006) Geological structure of a long-living geothermal system, Lardarello, Italy. Terra Nova 18:163-169

7. BGR (2012a) GeneSys Horstberg. http://www.genesys-hannover.de/Genesys/EN/Horstberg/horstberg_node_en.html Accessed 17 Sept. 2013
8. BGR (2012b) Milestones of the Genesys project. http://www.bgr.bund.de/Genesys/EN/Meilensteine/meilensteine_inhalt_en.html. Accessed 24 Sept 2012
9. BGR (2013) GeneSys Project Aktuelles.http://www.genesys-hannover.de/Genesys/DE/Aktuelles/aktuelles_node.html Accessed 12 Jan 2013
10. Bine 4 (2012) Korrosion und Materialqualifizierung. http://www.bine.info/service/bestellen/download-print/publikation/korrosion-in-geothermischen-anlagen/korrosion-und-materialqualifizierung/ . Accessed 18 Sept. 2013
11. BINE (2012a) Korrosion in geothermischen Anlagen. http://www.bine.info/service/bestellen/download-print/publikation/korrosion-in-geothermischen-anlagen/korrosion-und-materialqualifizierung/ . Accessed 18 Sept. 2013
12. BINE (2012b) Insheim - Geothermieanlage mit neuem Konzept. http://www.bine.info/newsuebersicht /news/insheim -geothermieanlage-mit-neuem-konzept. Accessed 05 Sept 2012
13. BINE (2012c) Geothermische Stromerzeugung im Verbund mit Wärmenetz. http://www.bine.info/service/bestellen/download-print/publikation/geothermische-stromerzeugung-im-verbund-mit-waermenetz/Accessed 21 Sept 2012
14. BINE (2012d) Projektinfo 14/07: Geothermische Stromerzeugung in Landau. http://www.bine.info /hau-ptnav igation/publikationen/publikation/geothermische-strome-rzeugung-in-landau/ . Accessed 20 Sept 2012
15. BLM (Bureau of Land Management) (2012) Newberry geothermal exploration project. http://www.blm.gov/or/districts/prineville/plans/newberry/index. php . Acc-essed 20 Sept 2012

16. Bloomquist R (2012) Integrating small power plants into agricultural projects. pangea.stanford.edu/ERE/pdf/IGAstandard/EGC/szeged/I-8-01.pdf. Accessed 01 Nov 2012
17. Blöscher G, Zimmermann G, Moeck I, Huenges E (2012) Groß-Schönebeck (D) – the development of the learning curve: experience from the projects of recent years. Oral presentation presented at ICEGS 2012, Freiburg. 25 May 2012
18. BMU (2011) Tiefe Geothermie - Nutzungsmöglichkeiten in Deutschland. Beltz Bad Langensalza GmbH, BT Weimar.
19. Bommer JJ, Oates S, Cepeda JM, Lindholm C, Bird J, Torres R, Marroquín G, Rivas J (2006) Control of hazard due to seismicity induced by a hot fractured rock geothermal project. Engineering Geology 83:287-306
20. Bracke R (2012) Geothermal energy – low enthalpy technologies. Oral presentation presented at Congreso Nacional de Energia 2012. CICR, San Jose/Costa Rica. 15–16 Feb 2012
21. Bromley C (2012) Geothermal induced seismicity: summary of international experience. Oral presentation presented at IEA-GIA Environmental Mitigation Workshop 2012, Taupo.15–16 June 2012
22. Bromley CJ, Mongillo MA (2008) Geothermal energy from fractured reservoirs - dealing with induced seismicity. IEA OPEN Energy Technology Bulletin, Issue No. 48. IEA. http://www.iea.org/impagr/cip/pdf/Issue48Geothermal.pdf. Accessed 31 May 2013
23. Brown D (1995) The US Hot Dry Rock program - 20 years of experience in reservoir testing. Paper presented at world geothermal congress 1995, Firenze.18–31 May 1995
24. Brown D (1997) Review of Fenton Hill HDR test results. Paper presented at New Energy and Industrial Technology Development Organization (NEDO) geothermal and HRD conference 1997, Sendai.10–17 Mar 1997
25. Brown D (2009) Hot Dry Rock Geothermal Energy: important lessons from Fenton Hill. Paper presented at thirty-fourth

workshop on geothermal reservoir engineering, Stanford University, Stanford.9–11 Feb 2009

26. Cappetti G (2006) How EGS is investigated in the case of the Lardarello geothermal field. In: Conference abstracts of engine launching conference. BRGM, Orleans.
27. Chabora E, Zemach E (2013) Desert Peak EGS Project. Geothermal Technologies office 2013 Peer Review.
28. Chabora E, Zemach E, Spielman P, Drakos P, Hickman S, Lutz S, Boyle K, Falconer A, Robertson-Tait A, Davatzes NC, Rose P, Majer E, Jarpe S (2012) Hydraulic stimulation of well 27–15, Desert Peak geothermal field, Nevada, USA. In: Proceedings of thirty-seventh workshop on geothermal reservoir engineering, Stanford University, Stanford.30 Jan–1 Feb 2012
29. Cherubini Y, Cacace M, Scheck-Wenderoth M, Moeck I, Lewerenz B (2013) Controls on the deep thermal field: implications from 3-D numerical simulations for the geothermal research site Groß Schönebeck. Environ Earth Sci Special.doi:10.1007/s12665-013-2519-4
30. Cladouhos TT, Osborn WL, Petty S, Bour D, Iovenitti J, Callahan O, Nordin Y, Perry D, Stern PL (2012) Newberry volcano EGS demonstration – phase I results. In: Proceedings of thirty-seventh workshop on geothermal reservoir engineering. Stanford University, Stanford.30 Jan–1 Feb 201
31. Clauser C (2006) Geothermal energy. In: Heinloth K (ed) Landolt-Börnstein group VIII: advanced materials and technologies, vol. 3: energy technologies, subvol. C: renewable energies, Springer, Heidelberg.
32. Cornet FH (2012) The learning curve: the Le Mayet de Montagne experiment (1978–1987). Oral presentation presented at ICEGS 2012, Freiburg.25 May 2012
33. Cotler S (2009) Enhanced geothermal energy project halted in the Geysers.http://stevecotler.com/tales/2009/09/03/egs-geysers-halted. Accessed 26 Sept 2012

34. Cummings RG, Morris GE (1979) Economic modeling of electricity production from Hot Dry Rock geothermal reservoirs: methodology and analysis. EA-630, Research Project 1017 LASL (LA-7888-HDR). OSTI Information Bridge. http://www.osti.gov/bridge/servlets/purl/5716131-wg4gUV/native/5716131.pdf. Accessed 31 May 2013
35. Davatzes N, Hickmann S, Zemach E, Spielman P, Robertson-Tait A, Drakos P, Lutz S, Rose P, Moore J, Majer E, Kennedy M, Stacey R, Swyer M (2012) Structural and geomechanical constraints in designing an EGS: example at Desert Peak Geothermal Field, Nevada. Oral presentation presented at ICEGS 2012, Freiburg.25 May 2012
36. DiPippo R (2012) Geothermal power plants. Elsevier, New York. pp 451-456
37. DiPippo R (2012) Geothermal power plants. Elsevier, New York. pp 463-474
38. Duchane D (1998) The history of HDR research and development. In: Draft proceedings of the 4th international HDR forum, Strasbourg.28–30 Sept 1998
39. Dumas P (2010) NER300: what for geothermal?. Oral presentation held at second EGEC TP geoelec meeting, Brussels.24 Mar 2010
40. EGEC (European Geothermal Energy Council) (2012) Geothermal market report, 2nd edition. European Geothermal Energy Council, Brussels.December 2012
41. EGEC (European Geothermal Energy Council) (2013) Fact sheet on enhanced geothermal systems: why it is different to shale gas. Accessed via the EGEC News, Issue No. 35.June 2013
42. EGS Coso (2013) Coso/EGS program.http://egs.egi.utah.edu/indexcoso.htm. Accessed 28 May 2013
43. Enbw (2013) Energie, die aus der Tiefe kommt – Das Geothermiekraftwerk in Bruchsal. http://www.enbw.com/content/de/der_konzern/_media/pdf/Flyer_Geothermie-V4silber-neu-grafik.pdf Accessed 30 Jan 2013

44. ENGINE (2008) ENGINE – geothermal lighthouse projects in Europe – Lardarello. http://engine.brgm.fr/mediapages/lighthouseProjects/LH-Quest_HydroTherm_2_Larderello.pdf. Accessed 17 Sept. 2013
45. ENGINE (2008b) ENGINE coordination action. Best practice handbook for the development of unconventional geothermal resources with a focus on enhanced geothermal system. BRGM, Orleans. http://engine.brgm.fr/Documents/ENGINE_BestPracticeHandbook.pdf . Accessed 25 Jan 2013
46. ENGINE (2012) ENGINE- geothermal lighthouse projects in Europe. http://engine.brgm.fr/mediapages/lighthouseProjects/LH-Quest_EGS_4_GeneSys.pdf. Accessed 24 Sept 2012
47. Evans K (2011) Enhanced/engineered geothermal systems: -experiences to date and lessons learned. In: Abstract volume of 9th Swiss geoscience meeting. ETH Zürich, Zürich.11–13 Nov 2011
48. Exorka (2013) F&E-Projekt Geothermie Allgäu 2.0.http://cif-ev.de/pdf/block2/P1.pdf Accessed 22 May 2013
49. Feige S, Roloff R (2012) Geothermal energy production – a subject for radiation protection. Paper presented at EUROSAFE Forum 2012, Brussels. 5–6 Nov 2012
50. Feldbusch E, Regenspurg S, Banks J, Milsch H, Saadat A (2013) Alteration of fluid properties during the initial operation of a geothermal plant: results from in situ measurements in Groß Schönebeck. Environ Earth Sci. doi:10.1007/s12665-013-2409-9
51. Fittermann DV (1988) Overview of the structure and geothermal potential of Newberry Volcano, Oregon. J Geophys Res 93:10059-10066
52. Garcia J, Walters M, Beall J, Hartline C, Pingol A, Pistone S, Wright M (2012) Overview of the Northwest Geysers EGS Demonstration Project. In: Proceedings of the thirty-seventh workshop on geothermal reservoir engineering (ed) Proceedings of the thirty-seventh workshop on geothermal

reservoir engineering, Stanford University, Stanford. 30 Jan–1 Feb 2012

53. Gebo NDS (Forschungsverbund Geothermie und Hochleistungsbohrtechnik) (2012a) Current state of research. http://www.gebo-nds.de/en/research-project/current-state-of-research/ . Accessed 12 Sept 2012
54. Gebo NDS (Forschungsverbund Geothermie und Hochleistungsbohrtechnik) (2012b) Research project. http://www.gebo-nds.de/en/research-project/. Accessed 20 Sept 2012
55. GeneSys (2012) GeneSys Hannover. http://www.genesys-hannover.de/Genesys/DE/Hannover/hannover_node.html. Accessed 3 June 2013
56. Genter A (2012) Lessons learned from projects in the early stage of EGS development: Soultz-sous-Fôrets (F). Oral presentation held at ICEGS 2012, Freiburg. 25 May 2012
57. Geodynamics (2013) Innamincka Deeps (EGS) project. http://www.geodynamics.com.au/Our-Projects/Innamincka-Deeps.aspx. Accessed: 18 Sept. 2013
58. Geothermie Stadt St. Gallen (2013a) Das Geothermie-Projekt der Stadt St. Gallen. http://www.geothermie.stadt.sg.ch/projekt.html. Accessed 21 May 2013
59. Geothermie Stadt St. Gallen (2013b) Das Geothermie-Projekt geht weiter.http://www.geothermie.stadt.sg.ch/aktuell/details/artikel/das-geothermie-projekt-geht-weiter.html. Accessed 29 August 2013
60. Geothermie-Nachrichten (2012) Pfalz: Geothermiekraftwerk in Insheim geht ans Netz.http://www.geothermie-nachrichten.de/pfalz-geothermiekraftwerk-in-insheim-geht-ans-netz. Accessed 16 Nov 2012
61. Geothermie-Pfalz (2013) Geothermieprojekt Insheim. http://geothermie-pfalz.de/geoinsh.html . Accessed 26 Aug 2013
62. Giardini D (2009) Geothermal quake risks must be faced. Nature 462:848-849

63. Goldstein B, Hiriart G, Bertani R, Bromley C, Gutiérrez-Negrín L, Huenges E, Muraoka H, Ragnarsson A, Tester J, Zui V (2011) Geothermal Energy. In: Edenhofer O, Pichs-Madruga R, Sokona Y, Seyboth K, Matschoss P, Kadner S, Zwickel T, Eickemeier P, Hansen G, Schlomer S, von Stechow C (eds) IPCC special report on renewable energy sources and climate change mitigation, Cambridge University Press, Cambridge. p 406
64. Grant MA, Garg SK (2012) Recovery factor for EGS. In: Proceedings of the 37th workshop on geothermal reservoir engineering. Stanford University, Stanford. 30 Jan–1 Feb 2012
65. Groos JC, Grund M, Ritter JRR (2012) Automated detection of microseismic events in the Upper Rhine valley near the city of Landau/South Palatinate. Geophys Res Abstracts 14:EGU2012-10482 2012, EGU General Assembly 2012
66. GtV (Bundesverband Geothermie) (2013) Liste der tiefen Geothermieprojekte in Deutschland. Liste der tiefen Geothermieprojekte in Deutschland, Liste der tiefen Geothermieprojekte in Deutschland. http://www.geothermie.de/wissenswelt/geothermie/in-deutschland.html. Accessed 20 Aug 2013
67. Häring MO (2007) Geothermische Stromproduktion aus Enhanced Geothermal Systems (EGS) Stand der Technik. http://www.geothermal.ch/fileadmin/docs/downloads/egs061207.pdf . Accessed 30 Jan 2013
68. Häring MO, Hopkirk R (2002) The Swiss deep heat mining project - the Basel exploration drilling. GHC Bulletin 23(1):31-33
69. Henninges J, Brandt W, Erbas K, Moeck I, Saadat A, Reinsch T, Zimmermann G (2012) Downhole monitoring during hydraulic experiments at the in-situ geothermal lab Groß Schönebeck. In: Proceedings of the thirty-seventh workshop on geothermal reservoir engineering. Stanford University, Stanford. 30 Jan 1–Feb 2012

70. Holl H (2012) Geodynamics Update: Innamincka Deeps EGS Project. Oral presentation held at ICEGS 2012, Freiburg. 25 May 2012
71. Huenges E (2010) Geothermal energy systems – exploration, development, and utilization. Wiley, Weinheim.
72. Huenges E (2013) Personal communication on 18th of January.
73. Ilyasov M, Ostermann I, Punzi A (2010) Modeling deep geothermal reservoirs: recent advances and future problems. In: Freeden W, Nashed MZ, Sonar T (eds) Handbook of geomathematics, vol. 1: general issues, key technologies, data acquisition, modeling the system earth, Springer, Berlin. pp 689-694
74. Informationsportal Tiefe Geothermie (2012) Beginn der Stimulation im EGS-Projekt Newberry Geothermal. http://www.tiefegeothermie.de/top-themen/beginn-der-stimulation-im-egs-projekt-newberry-geothermal. Accessed 21 May 2013
75. Informationsportal Tiefe Geothermie (2013a) Mauerstetten soll als Forschungsprojekt neu erschlossen werden. http://www.tiefegeothermie.de/news/mauerstetten-soll-als-forschungsprojekt-neu-erschlossen-werden. Accessed 21 May 2013
76. Informationsportal Tiefe Geothermie (2013b) Arbeitsmedien zur Stromerzeugung in Geothermiekraftwerken. http://www.tiefegeothermie.de/top-themen/arbeitsmedien-zur-stromerzeugung-in-geothermiekraftwerken Accessed 21 May 2013
77. Insheim (2012) Insheim. http://www.insheim.de/wirtschaft/geothermie.html. Accessed 9 May 2012
78. Johansson TB, Kelly H, Reddy AK, Williams RH (eds) (1993) Renewable energy: sources for fuels and electricity 2nd edn. Island Press, Washington D.C.
79. Julian BR, Foulger GR, Monastero FC (2009) Seismic monitoring of EGS stimulation tests at the Coso Geothermal

Field, California, using microearthquake locations and moment tensors. In: Proceedings of the thirty-fourth workshop on geothermal reservoir engineering. Stanford University, Stanford.

80. Jung R (2013) EGS–goodbye or back to the future, effective and sustainable hydraulic fracturing. In: Jeffrey R (ed) InTech, doi:10.5772/56458, http://www.intechopen.com/books/effective-and-sustainable-hydraulic-fracturing/egs-goodbye-or-back-to-the-future-95 . ISBN 978-953-51-1137-5. Accessed: 18 Sept 2013

81. Jupe AJ, Green ASP, Wallroth T (1992) Induced microseismicity and reservoir growth at the Fjällbacka hot dry rocks project, Sweden. Int J Rock Mechanics Mining Sci Geomechanics Abstracts 29(4):343-354

82. Kaieda H, Ito H, Kiho K, Suzuki K, Suenaga H, Shin K (2005) Review of the Ogachi HDR project in Japan. In: Proceedings of the world geothermal congress 2005, Antalya. 24–29 April 2005

83. Kaieda H, Sasaki S, Wyborn D (2010) Comparison of characteristics of micro-earthquakes observed during hydraulic stimulation operations in Ogachi, Hijiori and Cooper Basin HDR projects. In: Proceedings of the World Geothermal Congress 2010, Bali. 25–29 April 2010

84. Kappelmeyer O, Jung R (1987) HDR experiments at Falkenberg/Bavaria. Geothermics 16:375-392

85. Karner SL (2005) Stimulation techniques used in enhanced geothermal systems: perspectives from geomechanics and rock physics. In: Proceedings of the thirtieth workshop on geothermal reservoir engineering. Stanford University, Stanford. 31 Jan–2 Feb 2005

86. KIT (2013) Langzeitbetrieb und Optimierung eines Geothermiekraftwerks in einem geklüftet-porösen Reservoir im Oberrheingraben (LOGRO). http://www.agw.kit.edu/908_998.php Accessed 20 Sept 2012

87. Kreuter H (2011) Deep geothermal projects in Germany - status and future development. http://www.renewablesb2b.com/data/shared/GEO_PPT_Kreuter.pdf. Accessed 28 Sept 2012
88. Lacirignola M, Blanc I (2012) Environmental analysis of practical design options for enhanced geothermal systems (EGS) through life-cycle assessment. Renew Energy 50:901-914
89. Ladner F, Häring MO (2009) Hydraulic characteristics of the basel 1 enhanced geothermal system. GRC Transactions 33:199-203
90. Lazzarotto A, Sabatelli F (2005) Technological developments in deep drilling in the Lardarello area. In: Proceedings of the world geothermal congress 2005, Antalya. 24–29 April 2005
91. LGB-rlp (2012) (Landesamt für Geologie und Bergbau Rheinland-Pfalz) Steckbrief: Projektstandort Insheim. http://www.lgb-rlp.de/projektstandort_insheim.html . Accessed 06 Sept 2012
92. Lutz SJ, Schriener A Jr, Schochet D, Robertson-Tait A (2003) Geologic characterization of pre-tertiary rocks at the Desert Peak East EGS project site, Churchill County, Nevada. Geoth Res T 27:865-870 12–15 Oct 2003
93. Majer E, Baria R, Stark M, Oates S, Bommer J, Smith B, Asanuma H (2007) Induced seismicity associated with enhanced geothermal systems. Geothermics 36:185-222
94. Majer E, Nelson J, Robertson-Tait A, Savy J, Wong I (2011) Protocol for addressing induced seismicity associated with enhanced geothermal systems. http://www1.eere.energy.gov/geothermal/pdfs/geothermal_seismicity_protocol_012012.pdf. Accessed 05 June 2013
95. Majer E, Nelson J, Robertson-Tait A, Savy J, Wong I (2013) Best practices for addressing induced seismicity associated with enhanced geothermal systems (EGS), DRAFT, 23 May 2013. http://esd.lbl.gov/files/research/projects/induced_

seismicity/egs/Best_Practices_EGS_Induced_Seismicity_Draft_May_23_2013.pdf . Accessed 29 Aug 2013

96. Tester JW, Anderson BJ, Batchelor AS, Blackwell DD, DiPippo R, Drake EM, Garnish J, Livesay B, Moore MC, Nichols K, Petty S, Toksöz MN, Veatch RW Jr, MIT (2006a) The Future of geothermal energy - impact of enhanced geothermal systems on the United States in the 21st Century. US Department of Energy, Washington, D.C.. pp 1-10

97. Tester JW, Anderson BJ, Batchelor AS, Blackwell DD, DiPippo R, Drake EM, Garnish J, Livesay B, Moore MC, Nichols K, Petty S, Toksöz MN, Veatch RW Jr, MIT (2006b) The future of geothermal energy – impact of enhanced geothermal systems on the United States in the 21st Century. US Department of Energy, Washington, D.C.. pp 4-40

98. Tester JW, Anderson BJ, Batchelor AS, Blackwell DD, DiPippo R, Drake EM, Garnish J, Livesay B, Moore MC, Nichols K, Petty S, Toksöz MN, Veatch RW Jr, MIT (2006c) The future of geothermal energy – impact of enhanced geothermal systems on the United States in the 21st Century. US Department of Energy, Washington, D.C.. p 4

99. Tester JW, Anderson BJ, Batchelor AS, Blackwell DD, DiPippo R, Drake EM, Garnish J, Livesay B, Moore MC, Nichols K, Petty S, Toksöz MN, Veatch RW Jr, MIT (2006d) The future of geothermal energy – impact of enhanced geothermal systems on the United States in the 21st Century. US Department of Energy, Washington, D.C.. pp 4-26

100. Tester JW, Anderson BJ, Batchelor AS, Blackwell DD, DiPippo R, Drake EM, Garnish J, Livesay B, Moore MC, Nichols K, Petty S, Toksöz MN, Veatch RW Jr, MIT (2006e) The future of geothermal energy - impact of enhanced geothermal systems on the United States in the 21st Century. US Department of Energy, Washington, D.C.. pp 4-37

101. Tester JW, Anderson BJ, Batchelor AS, Blackwell DD, DiPippo R, Drake EM, Garnish J, Livesay B, Moore MC, Nichols K, Petty S, Toksöz MN, Veatch RW Jr, MIT (2006f) The future of

geothermal energy - impact of enhanced geothermal systems on the United States in the 21st Century. US Department of Energy, Washington, D.C.. pp 4-14 4–18

102. Tester JW, Anderson BJ, Batchelor AS, Blackwell DD, DiPippo R, Drake EM, Garnish J, Livesay B, Moore MC, Nichols K, Petty S, Toksöz MN, Veatch RW Jr, MIT (2006g) The future of geothermal energy - impact of enhanced geothermal systems on the United States in the 21st Century. US Department of Energy, Washington, D.C.. p 4–7–4–13
103. Tester JW, Anderson BJ, Batchelor AS, Blackwell DD, DiPippo R, Drake EM, Garnish J, Livesay B, Moore MC, Nichols K, Petty S, Toksöz MN, Veatch RW Jr, MIT (2006h) The future of geothermal energy - impact of enhanced geothermal systems on the United States in the 21st Century. US Department of Energy, Washington, D.C.. pp 4-38
104. Tester JW, Anderson BJ, Batchelor AS, Blackwell DD, DiPippo R, Drake EM, Garnish J, Livesay B, Moore MC, Nichols K, Petty S, Toksöz MN, Veatch RW Jr, MIT (2006i) The future of geothermal energy - impact of enhanced geothermal systems on the United States in the 21st Century. US Department of Energy, Washington, D.C.. p 4
105. Noack V, Scheck-Wenderoth M, Cacace M, Schneider M (2013) Influence of fluid flow on the regional thermal field: results from 3D numerical modelling for the area of Brandenburg (North German Basin). Environ Earth Sci. doi:10.1007/s12665-013-2438-4
106. Parker R (1999) The Rosemanowes HDR project 1983–1991. Geothermics 28:603-615
107. Pernecker G (1999) Altheim geothermal plant for electricity production by ORC-turbogenerator. In: Peter L (ed) Bulletin d'hydrogéologie No 17. Centre d'Hydrogéologie, Université de Neuchâtel, Altheim, Austria.
108. Petratherm (2012) Paralana. http://www.petratherm.com.au/projects/paralana . Accessed 19 Sept 2012

109. Pfalzwerke-geofuture (2012) Projekt Insheim. http://www.pfalzwerke-geofuture.de/6240.php . Accessed 05 Sept 2012
110. Pierce V (2011) Introduction to geothermal power. The English Press, Delhi.
111. Portier S, André L, Vuataz FD (2007) Review on chemical stimulation techniques in oil industry and applications to geothermal systems. Technical report in enhanced geothermal innovative network for Europe. CREGE - Centre for Geothermal Research, Neuchâtel.
112. Potter R, Robinson E, Smith M (1974) Method of extracting heat from dry geothermal reservoirs. US Patent No. 3,786,858, USA. Los Alamos, New Mexico. Accessed: 18 Sept 2013
113. Prevost JKJ (2004) The geothermal energy industry of El Salvador. Term Paper - ESD.166J Sustainable Energy, Spring 2004. web.mit.edu/10.391J/www/proceedings/Geothermal_Prevost2004.pdf. Accessed 26 Sept 2012
114. Regenspurg S, Blöcher G (2012) Saadat A (2012) Impact of geothermal stimulation treatment on the environment - a risk assessment. In: International conference on enhanced geothermal systems, Freiburg.
115. Rettenmaier D (2012) Lessons Learned - Reservoirmanagement Bruchsal. http://www.ta-survey.nl/pdf/GU2012-Detlev_Rettenmaier.pdf . Accessed 06 June 2013
116. Richards HG, Savage D, Andrews JN (1992) Granite-water reactions in an experimental hot dry rock geothermal reservoir, Rosemanowes test site, Cornwall, U.K. Appl Geochem 7(3):193-222
117. Rodríguez JA (2000) Geothermal development in El Salvador - a country update. In: Proceedings of the world geothermal congress 2000. Kyushu - Tohoku. 28 May–10 Jun 2000
118. Rodríguez JA (2003) Geothermal El Salvador. http://www.geothermal.org/articles/elsalvador.pdf . Accessed 26 Sept 2012
119. Rodríguez JA (2008) El Salvador Geothermal. http://www.

un.org/esa/sustdev/sids/2008_roundtable/presentation/energy_rodriguez.pdf. Accessed 26 Sept 2012

120. Romano B (2009) Geologist on trial in Basel: two EGS projects shuttered. http://www.rechargenews.com/energy/geothermal/article202104.ece . Accessed 26 Sept 2012
121. Romero A Jr, McEvilly TV, Majer E, Vasco D (1995) Characterization of the geothermal system beneath the Northwest Geysers Steam Field, California, from seismicity and velocity patterns. Geothermics 24:471-487
122. Rose P, Moore J, Kovac K, Adams M, McCulloch J, Spielman P, Sheridan J, Hickman S, Davatzes N, Julian B, Foulger G, Weidler R (2004) The Coso EGS Project - recent developments. Paper presented at Great Basin Geothermal Workshop, Reno. 5 Nov 2004
123. Sanjuan B, Jousset P, Pajot G, Debeglia N, De Michele M, Brach M, Dupont F, Braibant G, Lasne E, Duré F (2010) Monitoring of Bouillante geothermal exploitation (Guadeloupe, French West Indies) and the impact on its immediate environment. In: Proceedings of the world geothermal congress 2010, Bali. 25–30 April 2010
124. Sanyal SK, Butler SJ (2005) An analysis of power generation prospects from enhanced geothermal systems. In: Proceedings of the world geothermal congress 2005, Antalya. 24–29 April 2005
125. Sasaki S (1998) Characteristics of microseismic events induced during hydraulic fracturing experiments at the Hijiori hot dry rock geothermal energy site, Yamagata, Japan. Tectonophysics 289:171-188
126. Schanz U, Stang H, Tenzer H, Homeier G, Hase M, Baisch S, Weidler R, Macek A, Uhlig S (2003) Hot dry rock project Urach - a general overview. In: Proceedings of the European geothermal conference, Szeged. 25–30 May 2003
127. Schrage C, Bems C, Kreuter H, Hild S, Volland S (2012a) Overview of the enhanced geothermal energy project in Mauerstetten, Germany. http://www.ta-survey.nl/index.

php?id=109&lang=EN . Accessed 07 Sept 2012
128. Schrage C, Bems C, Kreuter H, Hild S, Volland S (2012b) Geothermie Allgäu 2.0 - overview of the enhanced geothermal energy project in Mauerstetten. Oral presentation held at Amsterdam, Germany.18 Apr 201
129. Schulte T, Zimmermann G, Vuataz F, Portier S, Tischner T, Junker R, Jatho R, Huenges E (2010) Enhancing geothermal reservoirs. In: Huenges E (ed) Geothermal energy systems, Wiley, Weinheim.
130. Stephens JC, Jiusto S (2010) Assessing innovation in emerging energy technologies: socio-technical dynamics of carbon capture and storage (CCS) and enhanced geothermal systems (EGS) in the USA. Energy Policy 38:2020-2031
131. Tenzer H (2001) Development of hot dry rock technology. GHC Bulletin, December.2001
132. Tenzer H, Schanz U, Homeier G (2000) HDR research programme and results of drill hole Urach 3 to depth of 4440m - the key for realisation of a HDR programme in Southern Germany and Northern Switzerland. In: Proceedings of the world geothermal congress. Kyushu, Tohoku. 25–30 April 2010
133. Tester JW, Brown DW, Potter RM (1989) Tester JW, Brown DW, Potter RM (1989) Hot dry rock geothermal energy - a new energy agenda for the 21st century. US Department of Energy, Washington D.C: Los Alamos National Laboratory report LA-11514-MS.
134. Tischner T, Evers H, Hauswirth H, Jatho R, Kosinowski M, Sulzbacher H (2010) New concepts for extracting geothermal energy from one well: the GeneSys-Project. In: Proceedings of the world geothermal congress, Bali. 25–30 April 2010
135. Wallroth T, Eliasson T, Sundquist U (1999) Hot dry rock research experiments at Fjällbacka, Sweden. Geothermics 28:617-625
136. Walters M (2013) Demonstration of an Enhanced Geothermal

System at the Northwest Geysers Geothermal Field, CA. Geothermal Technologies Office 2013 Peer Review.

137. Williams CF, Reed MJ, Anderson AF (2011) Updating the classification of geothermal resources. In: Proceedings of the thirty-sixth workshop on geothermal reservoir engineering. Stanford University, Stanford. 31 Jan–2 Feb 2011

138. Wyborn D (2011) Hydraulic stimulation of the Habanero enhanced geothermal system (EGS), South Australia. Presentation held at the 5th BC unconventional gas technical forum. April 2011 http://datafind.gov.bc.ca/query.html?qp=&style=ener&qt=mayet&Submit.x=0&Submit.y=0 Accessed 10 Sept 2012

139. Wyborn D, Graaf L, Davidson S, Hann S (2005) Development of Australia's first hot fractured rock (HFR) underground heat exchanger, Cooper Basin, South Australia. In: Proceedings of the world geothermal congress, Antalya. 24–29 April 2005

140. Zimmermann G, Tischner T, Legarth B, Huenges E (2009) Pressure-dependent production efficiency of an enhanced geothermal system (EGS): stimulation results and implications for hydraulic fracture treatments. Pure Appl Geophys 166:1089-1106

Chapter 3

Potential Water-related Environmental Risks of Hydraulic Fracturing Employed in Exploration and Exploitation of Unconventional Natural Gas Reservoirs in Germany

Axel Bergmann[1], Frank-Andreas Weber[1], H Georg Meiners[2], and Frank Müller[2]

[1]IWW Water Centre, Department Water Resources Management, Moritzstrasse 26, Muelheim 45476, Germany

[2]ahu AG Wasser Boden Geomatik, Kirberichshofer Weg 6, Aachen 52066, Germany

ABSTRACT

Background

The application of hydraulic fracturing during exploration and exploitation of unconventional natural gas reservoirs is currently under intense public discussion. On behalf of the German Federal Environment Agency we have investigated the potential water-related environmental risks for human health and the environment that could be caused by employing hydraulic fracturing in unconventional gas reservoirs in Germany. Here we provide an overview of the present situation and the state of the debate in Germany and summarize main results of the conducted risk assessment.

Results

We propose a concept for a risk assessment considering the site-specific analysis of the geosystem, the relevance of possible impact pathways and the hazard potential of the fracking fluids employed. The foundation of a sound risk analysis is a description of the current system, the relevant impact pathways and their interactions. An evaluation of fracking fluids used in Germany shows that several additives were employed even in newer fluids that exhibit critical properties or for which an assessment of their behaviour and effects in the environment is not possible or limited due to lack of current knowledge. The authors propose an assessment method that allows for the estimation of the hazard potential of specific fracking fluids, formation water, and the flowback based on legal thresholds and guidance values as well as on human- and eco-toxicologically predicted no-effect concentrations. The assessment of a previously employed and a prospectively planed fracking fluids shows that these fluids exhibit a high hazard potential. The flowback containing fracking fluid, formation water, and possibly reaction products can also exhibit serious hazard potentials, requiring environmentally acceptable techniques for its treatment and disposal.

Conclusions

The risk analysis must be conducted always site-specifically and consider regional groundwater flow conditions. The study concludes that currently missing knowledge and data prevent a profound assessment of the risks and their technical controllability in Germany. Missing knowledge and information includes data on the properties of the deep geosystem and of the behaviour and effects of the deployed chemical additives. In this setting the authors propose several recommendations for further action and procedures regarding the application of hydraulic fracturing in unconventional gas reservoirs in Germany.

BACKGROUND

The application of hydraulic fracturing ("fracking") in the exploration and exploitation of unconventional natural gas reservoirs has been generating intensive public debates in a variety of countries. Major concerns have focused on the potential impacts, hydraulic fracturing may cause on the environment and on human health, especially if fracking fluids contain toxic and environmentally harmful chemical additives.

Unconventional gas reservoirs are proven or presumed to be present in a number of different geological formations. An overview of potential geological host formations of unconventional gas reservoirs in Germany is given in Table 1, differentiating coalbed methane (CBM), shale gas and tight gas reservoirs. According to current estimates [1], the technologically recoverable gas reserves present in shale gas reservoirs in Germany amount to about 1,300 billion m^3 (estimates range from 0.7 to $2.3 \cdot 10^{12}$ m^3), assuming that 10% of the gas in place (GIP) is technologically recoverable. This estimated range of technologically recoverable shale gas reservoirs could, if exploited completely, cover the current annual gas consumption of Germany for 8 to 27 years [2]. The GIP in coalbed methane reservoirs in Germany is estimated to 450 billion m^3[3], but the technologically recoverable fraction has not yet been analysed. Conventional gas and tight gas reservoirs have been exploited in Germany over several decades, but current estimates of GIP remaining (100 billion $m^3$and 20 billion m^3, respectively [3]) indicate that the remaining reserves are limited.

Table 1: Potential unconventional gas reservoirs in Germany

Type of reservoir	Most promising reservoir	Regions
Coal bed methane (source rocks)	Seam-bearing Upper Carboniferous	Northern Ruhr region/Münsterland Basin (NRW)
		Ibbenbühen (NRW)
		Saar Basin (Saarland)
Shale gas (source rocks)	Tertiary clay formations (e.g. Fischschiefer)	Molasse Basin (BW)
	Posidonia Shale (Black Jurassic)*	Northwest German Basin (e.g. Lünne) (NI)
		Molasse Basin (BW)
		Upper Rhine Graben
	Wealden clay formations (e.g. Lower Cretaceous)*	Weser Depression (NRW/NI)
	Permian clay formations (e.g. black shale (stinkschiefe"), copper shale)	Northeast German Basin (NI/SA)
	Carboniferous and Devonian clay formations e.g. alum shale (Lower Carboniferous)*	Northern edge of the Rhenish massif (NRW)
		Northwest German Basin
		Harz Mountain (NI/SA)
	Silurian slates	Northeast German Basin
	Cambro Ordovician clay formations ("alum shale")	(not yet studied in details)

Tight gas (deposit rocks)	Red sandstone	Northwest German basin (NI/SA)
	Permian sandstones (Rotliegend) and carbonates (Zechstein)	Northeast German basin (e.g. Leer) (NI)
	Permian sandstones (Rotliegend) and dolomite (Stassfurt series) sandstones (Triassic)	Thuringian Basin (TH)
	Upper Carboniferous sandstones	Northwest German Basin (e.g. Vechta) (NI)

*indicates most relevant shale gas reservoirs according to [1].

The mining authorizations that have been issued for the exploration of unconventional gas reservoirs in Germany are shown in Figure 1. Most exploration has yet focused on the recovery and analysis of drilling core material as well as on geophysical methods, but hydraulic fracturing has already been applied in exploration at two sites [4]: at the site Damme 3 in Lower Saxony (3 fracs in the Wealden clay formation in depth of 1,045 – 1,530 m below ground surface using a slickwater fracking fluid in 2008) and at the site Natarp in North Rhine-Westphalia (2 fracs in CBM reservoirs in depth of 1,800 – 1,947 m using a gel fluid in 1995). To our knowledge, no mining authorizations have yet been approved for production-oriented exploitation of shale gas or CBM reservoirs in Germany. In the ongoing exploitation of tight gas and conventional gas reservoirs, however, experience in using hydraulic fracturing has been gained by pumping over 300 fracs over the last decades, mainly in the federal state of Lower Saxony [4]. In general, the exploited tight gas reservoirs in Germany are located in greater depth (often > 3.500 m) than some of the shale gas and CBM reservoirs currently considered for exploration, which vary in depth but are located partly in depth of 1.000 m or less[2,4-6], raising additional concerns on potential impacts on near-surface groundwater resources.

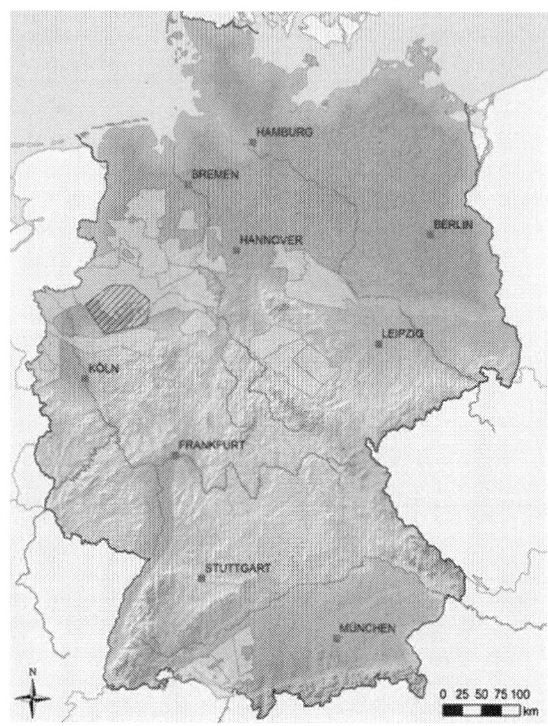

Figure 1: Mining authorizations in Germany (yellow, last revision: 31 December 2011) for exploration for unconventional hydrocarbon reservoirs (ochre: regions with the basic geological conditions for formation of shale gas) [1].

Driven by reports on the application and risk assessment of hydraulic fracturing in the U.S. [7-11], several risk assessments have recently been conducted on the specific German geological, technical, and legal situation, including an investigation on behalf of the German Federal Environment Agency (UBA) [4], a survey on behalf of the Ministry for Climate Protection, Environment, Agriculture, Nature Conservation and Consumer Protection (MKULNV) of the federal state of North Rhine-Westphalia [5], and an investigation of an independent expert group initiated by ExxonMobil Production Germany GmbH [12]. Given the current state of exploration of shale gas and CBM reservoirs in Germany, most risk assessments were conducted generically (i.e. not site-specific) or focused on some selected geological settings. Two site-specific investigations on regional situations in northern Hessian and in the river Ruhr watershed have recently been conducted [6,13].

Current State of the Debate in Germany

The political debate on hydraulic fracturing in Germany has proceeded as a result of the conducted risk assessments (or independent thereof), and new administrative procedures have been adapted recently.

The State Authority for Mining, Energy and Geology (LBEG) of Lower Saxony has issued minimum requirements for operating plans, criteria, and approval procedure for hydraulic treatments of boreholes in petroleum and natural gas reservoirs [14]. ExxonMobil Production Germany GmbH, a major operator in Germany, has announced that fracking projects in the vicinity of certain mineral spa protection zones are not further pursuit and no further hydraulic fracturing activities are carried out before suitable concepts for groundwater monitoring are implemented [15].

The state of North Rhine-Westphalia is currently not approving any exploration or production of natural gas from unconventional gas reservoirs, if harmful substances are employed [16]. A dialogue process is planned to involve the gas industry and communities, citizens, and relevant institutions in developing criteria for project approval and eliminating deficits of information and knowledge. In this context, borehole investigations, excluding hydraulic fracturing, are discussed for research purposes[17].

According to current press communications [18], the state of Lower Saxony is not approving further exploration and exploitation of shale gas and CBM reservoirs based on the lack of adequate risk assessment, but plans to continue approving exploitation of tight gas reservoirs in sandstone formation in depths > 2.500 m, as long as no environmentally toxic substances are injected underground. Draft legislations amending the environmental impact assessment

(EIA) regulation and of the Water Management Act (WHG) are currently discussed in Germany [19]. The drafts call for a ban of deep drillings involving hydraulic fracturing and the underground disposal of flowback in water protection zones, mineral spa protection zones, and in catchment areas of natural lakes from which raw water is procured directly for the public water supply. Based on the discussed draft legislation, the catchment area of artificial lakes and dams from which water is indirectly obtained for drinking water purposes would not generally be considered an exclusion zone [20].

Two regional investigations have analysed the regional occurrence of shale gas reservoirs in comparison to competing land-use obligations [6,13]. In a study on behalf of the river Ruhr water works consortium (*Arbeitsgemeinschaft der Wasserwerke an der Ruhr e.V.)* and the Ruhr River water board (*Ruhrverband*), we concluded that considering the regional occurrence of the shale gas reservoirs, the exclusion areas proposed by the draft legislations, and adopting criteria for the approval of exploitation involving hydraulic fracturing issued in Lower Saxony, an area of less than 3% of the issued mining authorization is accessible for exploitation of the shale gas reservoirs. Furthermore, a legal expertise commissioned by the Hessian Ministry of Environment, Energy, Agriculture and Consumer Protection (HLUG) has noted [13] that mining authorizations must not be granted if competing obligations among public stakeholders preclude subsequent exploitation of the gas reservoirs in the entire allocated field.

In the so-called "Hannover-Erklärung", the Federal Institute for Geosciences and Natural Resources (BGR), the Helmholtz Centre Potsdam - GFZ German Research Centre for Geosciences and the Helmholtz Centre for Environmental Research (UFZ) have called for the development of environmentally friendly fracking technology and proposed joint demonstration projects involving industry, research institutions, environmental organizations, and the general public [21]. An alliance of water suppliers, the Ruhr River water works consortium, and members of the beverage industry have called for clear legal provisions to protect the safety and purity of water resources from impacts of hydraulic fracturing in the so-called "Gelsenkirchener Erklärung" [22].

Furthermore, the exploitation of shale gas reservoirs is currently discussed controversial from an energy policy point of view. While the Federal Institute for Geosciences and Natural Resources (BGR) concluded that shale gas can contribute to domestic energy security [1], the German Advisory Council on the Environment (SRU) comes to the conclusion that the exploitation of shale gas using hydraulic fracturing is not necessary in Germany from an energy policy point of view and cannot substantially contribute to the transition to renewable energy sources [2].

Objectives

On behalf of the German Federal Environment Agency (UBA), a consortium of IWW Water Centre, ahu AG, [Gaßner, Groth, Siederer & Coll.], and Technical University of Darmstadt, has conducted a comprehensive investigation on potential environmental impacts of hydraulic fracturing related to exploration and exploitation of unconventional natural gas reservoirs, which focused on the framework of a risk assessment, the analysis of potential impact pathways, a method for assessing the hazard potentials of the fracking fluids employed, and on legal regulations and administrative structures [4]. Here we summarize main results of this study and propose recommendations for action and procedures. The study is based mostly on publicly accessible information including the relevant literature available internationally, but also on information provided by German authorities and operating companies.

RESULTS AND DISCUSSION

For assessing the risks that the application of hydraulic fracturing in unconventional natural gas reservoirs can pose on the water environment, we propose a concept that considers both the possible impact pathways and the potential hazard, any migration of the substances employed or encountered along these impact pathways could cause on exploitable water resources (Figure 2). Only if impact pathways are relevant for substance migration on the time scale considered, the substance-related hazard potentials cause adverse effects on the exploitable water environment. The risk of contamination of exploitable water resources is thus obtained by multiplying the relevance of the impact pathway(s) and the hazard potential of the pertinent fluids (fracking fluids and formation water). Since the state of knowledge does currently not allow for numerical calculations, we propose a five-part scale to evaluate the relevance of impact pathways and the hazard potential of the fluids involved (Figure 2).

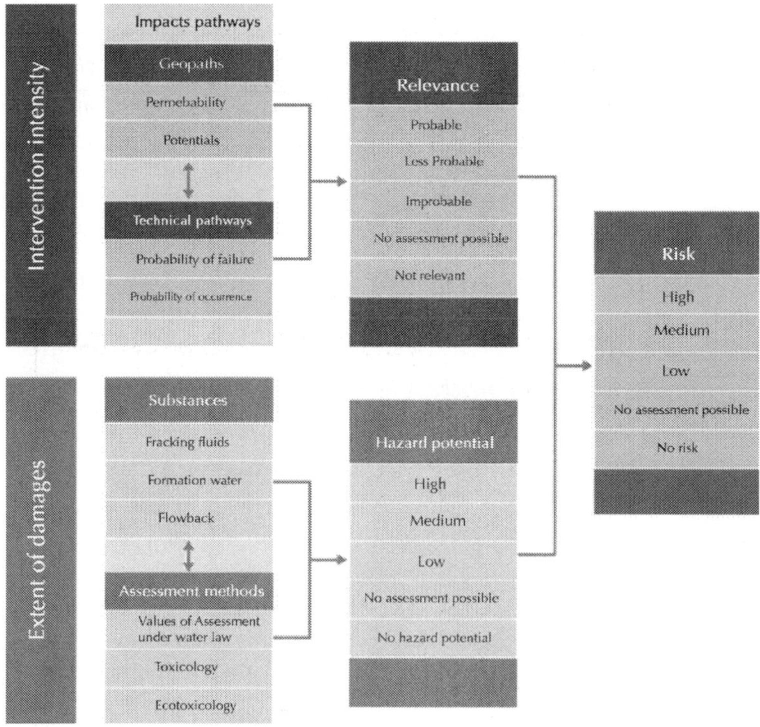

Figure 2: Structure of risk analysis for assessment of unconventional gas exploitation.

Impact Pathways

Potential water-related impact pathways are shown schematically in Figure 3, considering both technical and geological impact pathways. In most cases, failures of technical systems need to occur (such as failures of the well casing) for activating potential geological impact pathways (such as migration along faults), except in the fracking horizon, where no technical barrier is in place. Technical impact pathways could be quantified by probabilities of occurrence or probabilities of failure if data suitable for the German geological, technical and legal conditions were available. For a geological impact pathway to be relevant for substance migration, both permeability and hydraulic potentials must be considered for each geosystem site-specifically. Without suitable numerical quantification, however, the relevance of geological impact pathways can be estimated only with great uncertainties, for example using worst-case approaches.

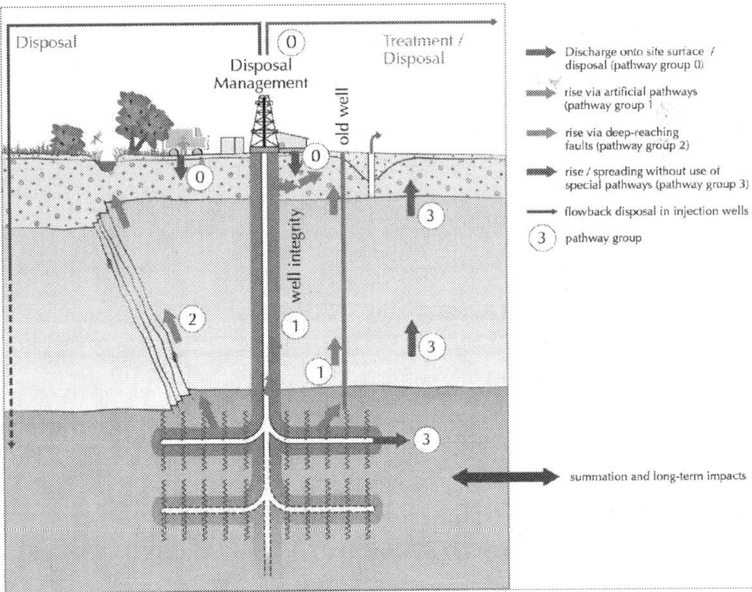

Figure 3: Schematics of potential impact pathways.

Pathway group 0 refers to (pollutant) discharges that occur directly at the ground surface, and especially in handling of fracking fluids (transport, storage, etc.) or flowback (e.g. via accidents or improper handling).

Pathway group 1 refers to potential (pollutant) discharges and migration along wells, i.e. to artificial underground pathways. With regard to the impact pathways involved, a distinction has to be made between production wells and old wells, such as wells from other explorations and uses.

Pathway group 2 comprises all impact pathways along geological faults. Significantly, the permeability along any given fault can vary, section-wise. Whereas deep-reaching, continuous faults can often be monitored, since the near-surface locations of their outcrops are usually known, faults that affect only parts of the overburden are difficult to monitor.

Pathway group 3 comprises extensive rise, as well as lateral spreading, of gases and fluids through geological strata (for example, via an aquifer), without preferred pathways similar to those described for pathway groups 1 and 2. Impact pathways in pathway group 3 depend primarily on the prevailing geological and hydrogeological conditions.

Summation and combination effects of the aforementioned impact pathways must be taken into account appropriately. Since many flow processes in the deep underground take place slowly, the relevant long-term impacts need to be considered. Such estimation is possible only on the basis of an extensive understanding of the geological and hydrogeological conditions prevailing in deep underground horizons, although not enough data of the studied geosystems are currently available to support conceptual or even numerical models.

Furthermore, the flowback disposal needs to be assessed as additional impact pathway, especially if flowback disposal is via injection into underground disposal wells.

Fracking Fluids

Overview

The fracking fluid is the hydraulic medium used for applying pressure to the rock strata inducing fracturing. With the fracking fluid, proppants (such as quartz sand) are transported into the created fractures in order to keep fractures from closing under the pressure of the surrounding rock and, thus, to ensure that the pathways created remain accessible for gas migration during the production phase. Fracking fluids usually contain a variety of chemical additives, with functions such as facilitating transport of proppants into fractures, preventing formation of precipitates, microbiological growth, formation of hydrogen sulphide, swelling of clay minerals, corrosion, and reducing fluid friction at high pump rates. Table 2 provides an overview of the functions of certain additives.

Table 2: Functions of additives used in fracking fluids (based on [4],[9])

Additive	Function
Proppants	Keeping the fractures created open under the pressure of the surrounding rock and allows gas/fluid to flow to the well bore
Scale inhibitors	Preventing deposits of poorly soluble precipitates, such as carbonates and sulphates
Biocides	Preventing bacterial growth, biofilm formation and formation of hydrogen sulphide by sulphate-reducing bacteria
Iron control	Preventing iron-oxide precipitation
Gelling agents	Improving proppant transport
High-temperature stabilizer (temperature stabilizer)	Preventing gel decomposition at high temperatures within the target horizon
Breakers	Reducing the viscosity of gel-containing fracking fluids for depositing proppants
Corrosion inhibitors	Protecting against equipment corrosion
Solvents	Improving the solubility of additives

pH regulators and buffers (pH control)	Controlling the pH of tracking fluids
Crosslinkers	Increasing viscosity at higher temperatures, to improve proppant transport
Friction reducers	Reducing friction within frac king fluids
Acids	Pretreating perforated sections of the well, and cleaning them of cement and drilling mud; dissolving acid-soluble minerals
Foams	Supporting proppant transport
H2S scavengers	Removing toxic hydrogen sulphide to protect equipment against corrosion
Surfactants	Reducing surface tension of fluids
Clay stabilizers	Reducing swelling and migration of clay minerals

Bergmann *et al.*

Bergmann *et al. Environmental Sciences Europe* 2014 26:10, doi:10.1186/2190-4715-26-10

In the following we present information on the fracking fluids and additives that have so far been employed in Germany. We then presented a method for assessing the hazard potentials of the fracking fluids employed with regard to groundwater, especially with regard to human use of groundwater as drinking water, and as part of natural cycles. In applying the method we assess selected fracking fluids used in Germany to date and possible new improvements of such fluids.

Fracking Fluids Used in Germany

We relied primarily on publicly accessible data to obtain information on the fracking fluids used in unconventional reservoirs in Germany [23]; only in some cases information from non-publicly accessible sources were obtainable [24]. The information on the composition of the fracking fluids used is based mainly on analyses of safety data sheets of the commercial products used to prepare fracking fluids. It has been found that these safety data sheets are often the only available source of information on the identity and the concentrations of the additives used. For approval authorities, this situation creates considerable uncertainties and lack of knowledge regarding the identity and the quantities of additives actually injected into the borehole.

Quantities Used

Information on fluid volumes was available for a total of 30 fracking fluids used in various unconventional reservoirs (and in one conventional reservoir) in Germany between 1982 and 2011. Most of the reservoirs in which the fluids were injected were tight gas reservoirs in Lower Saxony. The quantities used varied considerably, depending on the type of fracking fluid and the characteristics of the reservoirs. The quantities of fracking fluids used per frac ranged from less than 100 m^3 to more than 4,000 m^3. With the modern gel fluids used since 2000, an average of about 100 t of proppants and about 7.3 t of additives (of which usually less than 30 kg were biocides) were injected per frac. The quantities used can be quite large especially with multi-frac stimulations and/or use of slickwater fluids: for example, a total of about 12,000 m^3 of water, 588 t of proppants, and 20 t of additives (of which 460 kg were biocides) were injected into the "Damme 3" borehole in three frac operations in 2008.

Commercial Hydraulic Fracturing Products

According to the available information, at least 88 different hydraulic fracturing products have been used to prepare fracking fluids in Germany. However, since data are available on only 21 fracking fluids (corresponding to about 21% of the approximate 300 fracs carried out in Germany), it must be assumed that other products have also been employed. For 80 of the 88 products, we were able to obtain manufacturers' or importers' safety data sheets that were either current or valid at the time the fracs were carried out. Evaluation of the available 80 safety data sheets revealed that:

- 6 products are classified as toxic,
- 6 are classified as dangerous to the environment,

- 25 are classified as harmful,
- 14 are classified as irritant,
- 12 are classified as corrosive, and
- 27 are classified as non-hazardous

according to directives 67/548/EEC or 1999/45/EC, respectively. Several products are classified in more than one hazard class. With respect to the German water hazard classification (Wassergefährdungsklasse WGK), the commercial products were classified as follows according to the information in the safety data sheets:

- 3 preparations are classified as severely hazardous to waters,
- 12 preparations are classified as hazardous to waters,
- 22 preparations are classified as low hazardous to waters,
- 10 preparations are classified as not hazardous for water.

A total of 33 of the safety data sheets available to the study authors provided no information on the water hazard class of the product.

Fracking Additives

Information on the fracking additives used in the hydraulic fracturing products was available to the study authors for 28 fracking fluids. Those fluids were used in about 25% of 300 fracs carried out in Germany. Evaluation of those 28 fracking fluids showed that, overall, at least 112 substances/substance mixtures have so far been used in Germany. For 76 of the 112 substances/substance mixtures, either unique Chemical Abstracts Service (CAS) numbers were provided or it proved possible to correct or determine the CAS number on the basis of a unique given substance name. A total of 36 substances/substance mixtures could not be uniquely identified, either because their composition was unknown or because the available safety data sheets referred only to unspecific chemical group names (such as aromatic ketones, inorganic salts, etc.).

Hazard Potentials of Fracking Fluids

Comparison of two Fracking Fluids

Since recipes for fracking fluids are normally tailored to specific reservoirs, the hazard potentials of each fluid need to be assessed site-specifically. Based on the assessment method described in the Methods section, we have assessed the two fluids used to date in shale gas and CBM reservoirs in Germany as two examples. Planned improvements of fracking fluids were taken into account by assessing two fluids mentioned by an operator as potentially being suitable for shale gas reservoirs and, possibly, CBM reservoirs (improvements of slickwater and gel fluids) [4].

The hazard potentials of the slickwater fluid employed in the shale gas reservoir in 2008 and a planned improved composition are compared in Table 3. The assessment concludes that the slickwater fluid used in 2008 has a high toxicological and ecotoxicological hazard potential. In the improved fracking fluid, three hazardous additives that were still being used in 2008 are replaced by substances with considerably lower hazard potentials. However, also the improved fluid seems to exhibit a high hazard potential, because of employing high concentrations of a formaldehyde-forming biocide, for which little data is available for assessing its behaviour, fate, toxicity, and formation of degradation products. The replacement of the three hazardous additives that were still being used in 2008 by substances with considerably lower hazard potentials must be critically evaluated, since the underlying database for assessing those additives has been available for years, suggesting that service companies, operators, and/or authorities in the past have not always adequately considered the possibilities of substituting hazardous additives.

Table 3: Composition and hazard potential of two slickwater fluids

	Fracking fluid used at Damme 3				**Planned improvement of a slickwater fluid**			
Function	**Additive**	**Dissolved concentration in fracking fluid**	**Risk quotient based on toxicological assessment**	**Risk quotient based on eco-toxicological assessment**	**Additive**	**Dissolved concentration in fracking fluid**	**Risk quotient based on toxicological assessment**	**Risk quotient based on eco-toxicological assessment**
Clay stabilizer	Tetramethylammonium chloride	520 mg/L	1,733,000	Database insufficient (>2,600,000)	Cholinium chloride	750 mg/L	< 43	210
Friction reducer	Hydrotreated light petroleum distillates	220 mg/L	2,200	55,000	Butyl diglycol	350 mg/L	40	6,600
Surfactant	Ethoxylated octylphenol	36 mg/L	120,000	20,000	Polyethylene glycol monohexyl ether	130 mg/L	743	760
Biozide	Isothiazolinone derivative	4 mg/L	7,520	72,000	(Ethylenedioxy)-dimethanol	1,000 mg/L	10,000,000	Database insufficient (139,000)

Assessment of the fracking fluid used 2008 for hydraulic fracturing in a shale gas reservoir at Damme 3 and of a planned improvement based on human- and ecotoxicologically derived risk quotients.

Bergmann *et al.*

Bergmann *et al.* *Environmental Sciences Europe* 2014 26:10, doi:10.1186/2190-4715-26-10

Current developments aiming at reducing the numbers of additives used, at finding substitutes for substances that are highly toxic, carcinogenic, mutagenic, or toxic for reproduction, and at reducing or replacing biocidal agents, point to potential progress in the development of environmentally compatible fracking fluids. However, the authors can currently not evaluate the feasibility or progress of such efforts.

Flowback

Quantities and Composition

After pressure has been applied to the gas-bearing formation, some of the injected fracking fluids are recovered along with formation water and gas extracted from the well. The so-called flowback consists of varying proportions of injected fracking fluids and co-extracted formation water. Initially, fracking fluids account for the larger share of flowback; later, formation water predominates. As a result of various hydrogeochemical processes that can occur within the reservoir horizon (Figure 4), flowback can contain other substances in addition to fracking additives and formation water constituents.

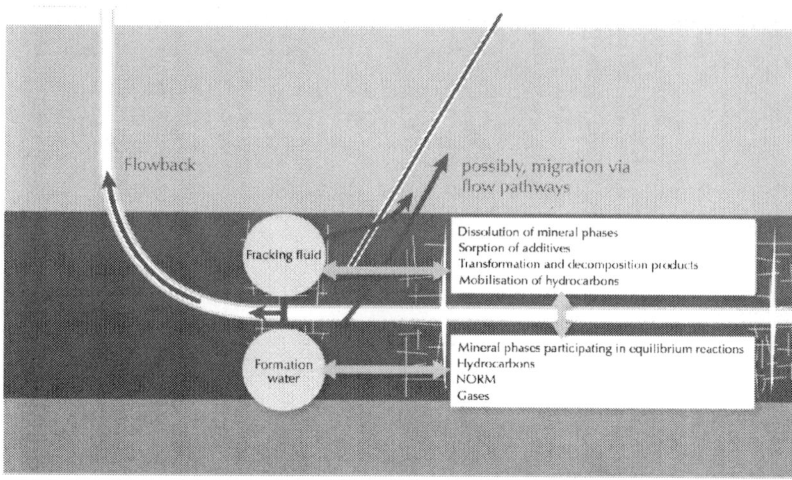

Figure 4: Hydrogeochemical processes affecting flowback formation via mixing of fracking fluids and formation water.

At the high pressures and temperatures prevailing in the target horizon, injected fracking additives may undergo chemical transformation and decomposition reactions in the presence of saline formation water. Microbiological decomposition reactions may occur as soon as the effects of the injected biocides diminish. In the process, metabolites can form that can pose toxicological and ecotoxicological risks potentially even exceeding the hazard posed by the parent substances that were injected.

Because the characteristics of formation water are always reservoir-specific, and because the proportions of extracted fracking additives vary, the characteristics of flowback have to be individually assessed for each site and pertinent time. Little information is available about the constituents of formation water in shale gas and CBM reservoirs in Germany, such as information about primary, secondary, and trace components, dissolved gases, organic substances, and NORM (Naturally Occurring Radioactive Material); regional and depth-oriented data is often missing.

At present, there is a lack of reliable analyses and mass balances that would allow for quantification of the variable mixing fractions, the fraction of the extracted fracking fluid, and possible reaction products. To date, no systematic measurements have been carried out for the purpose of identifying transformation and decomposition products in the flowback. Assessments of flowback from the "Damme 3" borehole carried out by Rosenwinkel et al. [25] concluded that only 8% of injected fracking fluids were being recovered as part of the flowback. Even though that percentage can be expected to increase as production continues, it seems certain that a substantial proportion of the fracking additives injected remains underground.

Disposal of Flowback

Possible technical processes for treating flowback have been reviewed by Rosenwinkel et al. [25] concluding that none of those treatment options, at present, qualifies as "best available technology" within the meaning of the German Federal Water Act. In general, the following options are possibly suitable for disposing or recycling of flowback in Germany:

- Underground injection via disposal wells,
- treatment for discharge into surface water,
- treatment for discharge into the sewer system,
- recycle and reuse in future hydraulic fracturing operations.

Operators currently refer to underground disposal of flowback as an important prerequisite for (cost-effective) exploitation of unconventional gas reservoirs. From the perspective of the study authors, flowback disposal via deep-underground injection can entail risks requiring site-specific risk assessment and monitoring.

CONCLUSIONS

There is general lack of basic information that would be needed for any well-founded assessment of the pertinent risks and the degree to which they can be controlled by technical means. Examples of such missing data include information regarding the structures and properties of deep geological systems (permeabilities, potential differences), the identities of the fracking additives used, and the chemical and toxicological properties of such additives. There are several reasons for this lack of information and data: (a) the information and data are not (openly) accessible, (b) the information and data have not yet been evaluated, and/or (c) there are gaps in knowledge that can only be closed through additional studies and research.

By studying selected geological systems in which shale gas or CBM reservoirs in Germany are found or assumed [4] we concluded that site-specifically certain impact pathways could be relevant for fluid migration. Little reliable data are currently available that would provide a basis for the reliable exclusion of risks to near-surface water resources. Assessment of selected fracking fluids used in unconventional gas reservoirs in Germany, along with the available information on the characteristics of flowback, have revealed that injected fluids, and fluids requiring disposal, can pose considerable hazard potentials. In summary, the study concludes that currently missing knowledge and data prevent a profound assessment of the risks and their technical controllability in Germany.

Recommendations

In light of the shortcomings of the currently available data, and of the fact that environmental risks cannot be ruled out, we recommend from a standpoint of precautionary water resources management, that above-ground and below-ground activities for unconventional gas exploitation involving fracking should not be approved for exploration or exploitation in water protection areas (classes I through III), water-extraction areas for the public drinking water supply (even if not assigned as water protection areas), mineral spa protection zones, and near mineral water reserves. These areas should be excluded for such activities. This recommendation on denial of approval should be reviewed as more data become available. In areas known to have unfavourable geological and hydrogeological conditions (groundwater potentials and known impact pathways), no exploration and exploitation of unconventional gas (via deep-drilling and hydraulic fracturing) should be allowed.

Site-specific risk assessment should be carried out with regard to any future drilling with fracking, and to drilling and use of underground disposal wells for injection of flowback. Such analyses should take account of all relevant fluids, whether introduced or encountered (fracking additives, formation water and its reaction products, and flowback), and of the relevant geological and technical impact pathways. It is recommended that use of toxicologically and ecotoxicologically hazardous fluids, and flowback disposal in disposal wells – also in the tight gas reservoirs in Germany that have already been exploited for many years – be reassessed.

Since the potential risks of exploration and exploitation of unconventional gas projects can be reliably assessed only if reliable information on the relevant geological systems (and potential impact pathways) is available, we recommend that any exploration of gas reservoirs provides investigations of the larger regional geological and hydrogeological system.

We further recommended that additional data and experience not yet published or not yet assessed (e.g., cadaster of old wells, cadaster of disposal wells) are evaluated and results are published. We argue however that without new data it will not be possible to answer the question of whether, and where, economically exploitable unconventional gas reserves are present in Germany and which technology (with or without fracking) is suited for exploration. We thus support the idea of carrying out further exploration, including exploration involving deep drilling (but without fracking), and carrying out targeted research in the above-described framework, for the purpose of answering those questions.

We recommend that further actions are taken step-by-step. Clear criteria should be established for deciding whether or not the application of fracking should be allowed at a later time. Such criteria should cover both the hazard potential of fracking additives and the availability of reliable information about the geological and technical impact pathways involved. Clear criteria should be applied for approval of any further exploration and any later production. A catalogue of criteria for approval should be developed step-by-step, applying transparent approaches involving public participation.

We recommend that research and development are intensified in areas such as the long-term integrity of wells, techniques available for forecasting the widths and lengths of fractures generated by fracking, and the development of fracking fluids with lower hazard potential. Practical application of the relevant research findings should be monitored scientifically.

With regard to EIA obligations, we recommend that fracking projects be subject to general federal EIA obligations, and that such obligations include an "opening clause" to allow participation of the German federal states. The public participation required under EIA legislation should be expanded to include a project-monitoring component, since many findings regarding projects' potential environmental impacts cannot be obtained until the projects are actually underway. Careful review of requirements under water law should be assured, via clarification of pertinent requirements, and via a) introduction of an integrated project-approval procedure to be directed by an environmental authority subordinate to the Ministry for the Environment, or b) integration of mining authorities within the environmental administration.

The following two aspects are of central importance with regard to any continuation of exploration and exploitation of unconventional gas in Germany, regardless of the procedures applied: all work processes and results should be fully transparent, and all stakeholders should exercise trust in their dealings with each other. Efforts should include the establishment of a publicly accessible cadaster listing all fracking measures carried out in the past, along with the quantities and the compositions of the fluids used.

In the following sections, we propose special recommendations for further steps towards exploitation of unconventional gas reservoirs in Germany. The focus of the recommendations is on the next phase of pilot exploration, especially, exploration in geological systems for which no information, or very little information, is yet available concerning unconventional gas reservoirs they may contain. The objectives of the recommendations include:

- identifying hydrogeological problematic areas, and possible impact pathways, at an early stage, and proposing measures for ongoing monitoring,
- reducing the hazard potential of the fracking fluid potentially used.

Special Recommendations With Regard to the Area of Geological Systems and the Aquatic Environment

The cause-and-effect relationships between deep-reaching and near-surface groundwater flow systems are of particular importance with regard to the water-related environmental impacts of unconventional gas exploitation projects. Such assessments require a detailed understanding of the hydrogeological systems involved, including:

- Conceptual hydrogeological models should be prepared that support reliable risk assessment for all potential impact pathways. The scope of such conceptual models should be large enough to support assessment of the impacts of exploration and exploitation of unconventional gas – via fracking – both for the specific sites and with regard to the large geological systems (system-oriented exploration).
- For areas in which water-related environmental impacts cannot be ruled out, numerical groundwater flow models should be prepared/refined in order to quantify the pertinent risks. This may involve preparing a regional model that can serve as a basis for local numerical models in the exploration area.
- The aforementioned numerical models have to be continually verified and calibrated on the basis of data and information obtained through monitoring (both prior and during the

project). For monitoring to be effective, it must be based on an adequate understanding of the system involved. At the same time, the understanding of the system involved (conceptual or numerical model) can be improved by the monitoring data obtained. Monitoring-based project control requires meaningful indicators and an evaluation system. Ultimately, options must be available for stopping, limiting, or reversing any undesired developments. The models resulting from the aforementioned work steps provide an important basis for authorities' decisions regarding the approval of submitted projects, as well as possible ancillary provisions under water law.

- The necessary regional and local models must be provided by the mining company within the authorization procedure under mining law and water law, based on the requirements imposed by the competent mining and water authorities. A fracking project can be approved only when enough pertinent knowledge has been gained and adequate precaution has been taken to exclude any adverse impact on exploitable water resources.

Special Recommendations with Regard to the Area of Substances

Assessment of selected fracking fluids used in unconventional gas reservoirs in Germany, along with the available information on the characteristics of flowback, have revealed that injected fluids, and fluids requiring disposal, can pose considerable hazard potentials. In light of the gaps in knowledge, uncertainties and data deficits identified via research and assessment for the present study, the following recommendations for action are seen as important:

- Complete disclosure of all substances used, with regard to substance identities and quantities.
- Assessment of the toxicological and ecotoxicological hazard potentials of substances used, and provision of all physical-

chemical and toxicological substance data required by the mining company. If relevant substance data are lacking, the gaps in the data must be eliminated – if necessary, via suitable laboratory tests or model calculations. In the process, the effects of relevant substance mixtures must be taken into account.

- Substitution of unsafe substances (especially, substances that are highly toxic, carcinogenic, mutagenic, or toxic for reproduction), reduction or substitution of biocides, reduction of the numbers of additives used, lowering of concentrations used.
- Determination and assessment of the characteristics of site-specific formation water, with regard to constituents of relevance to drinking water quality (salts, heavy metals, Naturally Occurring Radioactive Material – NORM, hydrocarbons).
- Determination and assessment of the characteristics of site-specific flowback, with regard to constituents of relevance to drinking water quality (salts, heavy metals, NORM, hydrocarbons), and with regard to additives used (primary substances) and their transformation products (secondary substances); determination and assessment of the proportion of fracking fluids recovered with the flowback.
- Determination of the behaviour and fate of substances in the fracking horizon, via mass balancing of the additives used.
- Modelling of substance transport, for assessment of possible risks to any exploitable groundwater, from any migrating formation water and fracking fluids.
- Technical treatment and "environmentally sound" disposal of flowback, including description of all technically feasible treatment processes and of the possibilities for reusing substances. If injecting flowback into disposal wells, conducting of a site-specific risk analysis is recommended.
- Monitoring and system-oriented examination, including installation of near-surface groundwater observation wells to

determine the reference condition with regard to additives and methane; if appropriate, installation of deep groundwater observation wells to determine the characteristics of formation water and the relevant hydraulic potentials.

METHODS

Under German water law, the key requirement to be applied in assessing releases of substances into the groundwater is that releases must not adversely affect the water quality (Art. 48 (1) WHG, Federal Water Resources Management Act). An adverse effect on the quality of near-surface groundwater (i.e. of the exploitable groundwater that is integrated in natural cycles) has occurred, if water quality has worsened more than slightly.

An adverse effect on the water quality of groundwater must be assumed if relevant legal and sub-legal limit values, guide values, maximum values, and especially the "Geringfügigkeitsschwellenwerte" (de minimis thresholds, GFS) of the German Federal/State Working Group on Water (LAWA) [26] are exceeded in any exploitable groundwater. These de minimis thresholds are primarily based either on the maximum permitted concentration specified by the Ordinance on Drinking Water (Trinkwasserverordnung), or, if no maximum permitted concentration has yet been established, on toxicologically and ecotoxicologically derived threshold values. Thus, it is ensured that groundwater remains available as drinking water resource for human consumption, and it remains intact as a habitat and as part of natural cycles.

For the majority of the substances used as fracking additives, no de minimis thresholds or other water-law-based assessment values have yet been established. Therefore, hygienic guidance values for drinking water (GVDW – maximum concentration of a substance in drinking water that can be tolerated for a lifetime without suffering adverse effects on health) or health orientation values (HOV - precautionary value for substances that cannot (or can only partially) be toxicologically assessed [27]) and ecotoxicologically established Predicted No Effect Concentrations (PNEC - maximum concentration of a substance at which no effects on organisms of an aquatic ecosystem are expected [28]) were assessed for such substances, or derived using published methods, following the concept of LAWA[26].

Relevant for the assessment is the concentration at the location where the substance enters exploitable groundwater resources. In case of substances entering groundwater from the surface (pathway group 0, e.g. accidents during transport and preparation of fracking fluids), the relevant substance concentration for the assessment is the concentration at the groundwater surface (see page water). By analogy, in the case of a possible release from the fracking horizon (and related migration via pathway groups 1 through 3), the concentration at the base of the exploitable groundwater aquifer should be used in the assessment. The relevant substance concentrations can properly assessed only site-specifically. For potential migration and exposure scenarios, suitable models are needed that consider relevant hydraulic and geochemical transport, mixing, decomposition, and reaction processes along the underground flow pathway. No such models are available at present that have the necessary spatial resolution.

As long as suitable models are lacking, we propose to assess hazard potentials on the basis of substance concentrations in (undiluted) fracking fluids and formation water. Based on the current state of knowledge, we consider it not suitable to presume a considerable reduction of their hazard potential due to dilution along the underground flow pathways, because along the flow path dilution occurs mainly by mixing with saline groundwater, which can have considerable hazard potential of its own (see below); thus, mixing with such water would not necessarily reduce the hazard potential of fracking fluids. The pertinent hazard potentials of the fluids are assessed on the basis of the individual constituents, calculating substance-specific risk quotients of substance concentrations and assessment values (GFS, GVDW, HOV, or PNEC):

$$\text{Risk Quotient} = \frac{\text{substance concentration in the fluid}}{\text{assessment value}}$$

When a substance has a risk quotient < 1, no hazard potential is expected, while a risk quotient ≥ 1 represents potentially a toxicological or ecotoxicological hazard (hazard potential). In the present study, a risk quotient $> 1{,}000$ is assumed to represent a high hazard potential. This value is given as an example and has not been scientifically established; it needs to be site-specifically reviewed on the basis of exposure scenarios – using numerical models for example.

AUTHORS' CONTRIBUTIONS

The authors contributed in equal parts to this publication. All authors read and approved the final manuscript.

ACKNOWLEDGEMENTS

The authors would like to thank the German Federal Environment Agency (UBA) for financing the study and the project partners [Gassner, Groth, Siederer & Coll.] and TU Darmstadt (Prof. Dr. Sass) for their collaboration.

REFERENCES

1. Bundesanstalt für Geowissenschaften und Rohstoffe (2012) Abschätzung des Erdgaspotenzials aus dichten Tongesteinen (Schiefergas) in Deutschland. Hannover. http://www.bgr.bund.de/DE/Themen/Energie/Downloads/BGR_Schiefergaspotenzial_in_Deutschland_2012.pdf?__blob=publicationFile&v=7

2. Sachverständigenrat für Umweltfragen: Fracking zur Schiefergasgewinnung (2013) Ein Beitrag zur energie- und umweltpolitischen Bewertung. Aktuelle Stellungnahme Nr. 18. Berlin. http://www.umweltrat.de/SharedDocs/Downloads/DE/04_Stellungnahmen/2012_2016/2013_05_AS_18_Fracking.pdf?__blob=publicationFile
3. Bundesanstalt für Geowissenschaften und Rohstoffe (2013) Energiestudie 2013. Reserven, Ressourcen und Verfügbarkeit von Energierohstoffen. Hannover. p 112 http://www.bgr.bund.de/DE/Themen/Energie/Downloads/Energiestudie_2013.pdf?__blob=publicationFile&v=5
4. Umweltbundesamt (2012) Umweltauswirkungen von Fracking bei der Aufsuchung und Gewinnung von Erdgas aus unkonventionellen Lagerstätten – Risikobewertung, Handlungsempfehlungen und Evaluierung bestehender rechtlicher Regelungen und Verwaltungsstrukturen. -Gutachten im Auftrag des Umweltbundesamtes. Berlin. http://www.umweltbundesamt.de/uba-info-medien/4346.html
5. Ministerium für Klimaschutz, Umwelt, Landwirtschaft, Natur- und Verbraucherschutz des Landes NRW (2012) Gutachten mit Risikostudie zur Exploration von Erdgas aus unkonventionellen Lagerstätten in Nordrhein-Westfalen und deren Auswirkungen auf den Naturhaushalt, insbesondere die öffentliche Trinkwassergewinnung. Düsseldorf. http://www.umwelt.nrw.de/umwelt/wasser/trinkwasser/erdgas_fracking
6. IWW Rheinisch-Westfälisches Institut für Wasser Beratungs- und Entwicklungsgesellschaft mbH (2013) Wasserwirtschaftliche Risiken bei Aufsuchung und Gewinnung von Erdgas aus unkonventionellen Lagerstätten im Einzugsgebiet der Ruhr. Gutachten des IWW im Auftrag der Arbeitsgemeinschaft der Wasserwerke an der Ruhr e.V. und des Ruhrverbandes. Mülheim. http://www.awwr.de/fileadmin/download/download_2013/studie_fracking_einzugsgebiet_ruhr.pdf , http://www.ruhrverband.de/wissen/forschung-entwicklung/fracking/

7. U.S. Environmental Protection Agency (2004) Evaluation of impacts to underground sources of drinking water by hydraulic fracturing of coalbed methane reservoirs. http://water.epa.gov/type/groundwater/uic/class2/hydraulicfracturing/wells_coalbedmethanestudy.cfm
8. U.S. Environmental Protection Agency (2011) Plan to Study the Potential Impacts of Hydraulic Fracturing on Drinking Water Resources. Washington. http://water.epa.gov/type/groundwater/uic/class2/hydraulicfracturing/upload/hf_study_plan_110211_final_508.pdf
9. Tyndall Centre (2011) Shale gas: a provisional assessment of climate change and environmental impacts. Manchester. http://www.tyndall.ac.uk/shalegasreport
10. Waxman HA, Markey EJ, Degette D (2011) Chemicals used in hydraulic fracturing. In: U.S. House of Representatives Committee on Energy and Commerce Minority Staff. Washington. http://democrats.energycommerce.house.gov/sites/default/files/documents/Hydraulic-Fracturing-Chemicals-2011-4-18.pdf
11. New York State Department of Environmental Conservation (2011) Revised Draft Supplemental Generic Environmental Impact Statement. Chapter 5: Natural gas development activities & high-volume hydraulic fracturing. New York. http://www.dec.ny.gov/docs/materials_minerals_pdf/rdsgeisch50911.pdf
12. Ewen C, Borchardt D, Richter S, Hammerbacher R (2012) RisikostudieFracking–ÜbersichtsfassungderStudie"Sicherheit und Umweltverträglichkeit der Fracking-Technologie für die Erdgasgewinnung aus unkonventionellen Quellen" erstellt im Zusammenhang mit dem InfoDialog Fracking. Darmstadt. http://dialog-erdgasundfrac.de/sites/dialog erdgasundfrac.de/files/Ex_risikostudiefracking_120518_webprint.pdf
13. Hessischer Landtag (2013) 60. Sitzung des Ausschusses für Umwelt, Energie, Landwirtschaft und Verbraucherschutz.

Wiesbaden. http://www.hessischer-landtag.de/icc/med/bb7/bb700690-9433-e31a-628b-31402184e373,

14. Landesamt für Bergbau, Energie und Geologie Niedersachsen Mindestanforderungen an Betriebspläne, Prüfkriterien und Genehmigungsablauf für hydraulische Bohrlochbehandlungen in Erdöl- und Erdgaslagerstätten in Niedersachsen. Clausthal-Zellerfeld. Rundverfügung vom 31.10.2012.http://www.lbeg.niedersachsen.de/download/72198/Mindestanforderungen_an_Betriebsplaene_Pruefkriterien_und_Genehmigungsablauf_fuer_hydraulische_Bohrlochbehandlungen_in_Erdoel-_und_Erdgaslagerstaetten_in_Niedersachsen.pdf
15. Hammerbacher Beratung & Projekte Statusbericht von ExxonMobil zur Umsetzung der Risikostudie Fracking. Osnabrück. Protokoll vom 6. November 2012, Osnabrück http://www.erdgassuche-in-deutschland.de/dialog/info_dialog_fracking_status.html
16. Ministerium für Klimaschutz, Umwelt, Landwirtschaft, Natur- und Verbraucherschutz des Landes Nordrhein-Westfalen Pressemitteilung vom 07.09.2012 - Umweltministerium und Wirtschaftsministerium legen Risikogutachten zu Fracking vor. http://www.umwelt.nrw.de/ministerium/service_kontakt/archiv/presse2012/presse120907_a.php
17. Remmel J (2012) Erdgas aus unkonventionellen Lagerstätten. gwf Wasser Abwasser 11:1121
18. Niedersächsisches Ministerium für Wirtschaft, Arbeit und Verkehr Gemeinsame Presseinformation von Minister Wenzel und Lies vom 17.03.2014 - Ja zur Erdgasförderung! Nein zu umwelttoxischen Substanzen unter Tage! http://www.mw.niedersachsen.de/portal/live.php?navigation_id=5459&article_id=123032&_psmand=18
19. Bundesministerium für Umwelt, Naturschutz und Reaktorsicherheit Vorschlag zur Änderung von UVP-V und Wasserhaushaltsgesetz. http://www.bmu.de/themen/wasser-

abfall-boden/binnengewaesser/gesetzesaenderung-zu-fracking

20. Deutscher Verein des Gas- und Wasserfaches e.V Stellungnahme vom 21. März 2013 zum Entwurf eines Gesetzes zur Änderung des Wasserhaushaltsgesetzes vom 7. März 2013 und Entwurf einer Verordnung zur Änderung der Verordnung über die Umweltverträglichkeitsprüfung bergbaulicher Vorhaben vom 11. März 2013 in Bezug auf die Umweltverträglichkeitsprüfung bei Bohrungen mit Einsatz der Fracking-Technologie. http://www.dvgw.de/wasser/ressourcenmanagement/gewaesserschutz/fracking/
21. BGR, GFZ & UFZ (2013) Abschlusserklärung zur Konferenz "Umweltverträgliches Fracking?". Hannover. am 24./25. Juni 2013 (Hannover-Erklärung).http://www.bgr.bund.de/DE/Gemeinsames/Nachrichten/Veranstaltungen/2013/GZH-Veranst/Fracking/Downloads/Hannover-Erklaerung-Finalfassung.pdf
22. Gelsenwasser AG, Arbeitsgemeinschaft der Wasserwerke an der Ruhr e.V., Deutscher Brauer–Bund e.V., Verband Deutscher Mineralbrunnen e.V. & Wirtschaftsvereinigung Alkoholfreie Getränke e.V (2013) Gelsenkirchener Erklärung: Wasserversorger, Bierbrauer, Mineral– und Heilbrunnenbetriebe sowie Erfrischungsgetränkehersteller fordern Schutz vor Fracking. Gelsenkirchen. (24.10.2013) http://www.gelsenwasser.de/fileadmin/download/unternehmen/presse/gelsenkirchener_erklaerung.pdf
23. ExxonMobil Central Europe Holding GmbH Frack-Flüssigkeiten. http://www.erdgassuche-in-deutschland.de/erkundung_foerderung/frac_fluessigkeiten/index.html
24. Bezirksregierung Arnsberg (2011) Gewinnung von Erdgas aus unkonventionellen Lagerstätten – Erkundungsmaßnahmen der CONOCO Mineralöl GmbH in den Jahren 1994 – 1997. Arnsberg. 61.01.25-2010-9
25. Rosenwinkel KH, Weichgrebe D, Olsson O (2012) Gutachten Stand der Technik und fortschrittliche Ansätze in der Entsorgung

des Flowback des Instituts für Siedlungswasserwirtschaft und Abfall (ISAH) der Leibniz-Universität Hannover zum Informations- und Dialogprozess über die Sicherheit und Umweltverträglichkeit der Fracking-Technologie für die Erdgasgewinnung. Hannover. http://dialog-erdgasundfrac.de/sites/dialog-erdgasundfrac.de/files/Gutachten%20zur%20Abwasserentsorgung%20und%20Stoffstrombilanz%20ISAH%20Mai%202012.pdf

26. LAWA – Bund/Länder-Arbeitsgemeinschaft Wasser Ableitung von Geringfügigkeitsschwellen für das Grundwasser. Düsseldorf: 2004. http://www.lawa.de/documents/GFS-Bericht-DE_a8c.pdf

27. Umweltbundesamt (2003) Bewertung der Anwesenheit teil- oder nicht bewertbarer Stoffe im Trinkwasser aus gesundheitlicher Sicht. Empfehlung des Umweltbundesamtes nach Anhörung der Trinkwasserkommission des Bundesministeriums für Gesundheit. Bundesgesundheitsbl Gesundheitsforsch Gesundheitsschutz 46:249-251

28. European Commission (2003) Technical Guidance Document in support of Commission Directive 93/67/EEC on Risk Assessment for new notified substances, Commission Regulation (EC) No 1488/94 on Risk Assessment for existing substances and Directive 98/9/EC of the European Parliament and of the Council concerning the placing of biocidal products on the market, Part II. Ispra. http://ihcp.jrc.ec.europa.eu/our_activities/public-health/risk_assessment_of_Biocides/doc/tgd

Chapter 4

Physical, Chemical, and Biological Characteristics of Compounds Used in Hydraulic Fracturing

William T. Stringfellow[a, b], Jeremy K. Domen[a], Mary Kay Camarillo[a], Whitney L. Sandelin[a], and Sharon Borglin[b]

[a]Ecological Engineering Research Program, School of Engineering & Computer Science, University of the Pacific, 3601 Pacific Avenue, Stockton, CA 95211, USA

[b]Earth Sciences Division, Lawrence Berkeley National Laboratory, 1 Cyclotron Road, Berkeley, CA 94720, USA

ABSTRACT

Hydraulic fracturing (HF), a method to enhance oil and gas production, has become increasingly common throughout the U.S. As such, it is important to characterize the chemicals found in HF fluids to evaluate potential environmental fate, including fate in treatment systems, and human health impacts. Eighty-one common HF chemical additives were identified and categorized according to their functions. Physical and chemical characteristics of these additives were determined using publicly available chemical information databases. Fifty-five of the compounds are organic and twenty-seven of these are considered readily or inherently biodegradable. Seventeen chemicals have high theoretical chemical oxygen demand and are used in concentrations that present potential treatment challenges. Most of the HF chemicals evaluated are non-toxic or of low toxicity and only three are classified as Category 2 oral toxins according to standards in the *Globally Harmonized System of Classification and Labeling of Chemicals*; however, toxicity information was not located for thirty of the HF chemicals evaluated. Volatilization is not expected to be a significant exposure pathway for most HF chemicals. Gaps in toxicity and other chemical properties suggest deficiencies in the current state of knowledge, highlighting the need for further assessment to understand potential issues associated with HF chemicals in the environment.

INTRODUCTION

Hydraulic fracturing (HF) is a technique where fluids are pumped into wells under high pressure (e.g. up to 69,000 kPa in Marcellus Shale [1]) in order to fracture low permeability geologic formations to increase formation permeability [2] and [3]. HF is commonly applied to increase permeability in shale, tight sands, coal-beds, and other gas and oil-bearing strata, resulting in higher oil and gas production [2]. HF is also used in conventional oil and gas reservoirs to enhance production, improving the life cycle of previously developed oil and gas sources [3]. In addition, HF is used to develop geothermal energy resources, exposing greater surface area for heat transfer between injected fluids and thermal formations [4]. HF is also used to enhance water production in "tight" geologic formations where water is being extracted.

Extraction of hydrocarbon resources from "tight" geologic formations by the use of HF is commonly referred to as "unconventional production". Unconventional production is increasing globally as a result of technological advances in horizontal drilling, improvements in fracturing technology, and market forces which collectively have made HF an economically competitive technology for resource recovery [5] and [6]. In the U.S., unconventional production of natural gas from shale is projected to increase to 385 billion m^3 yr^{-1} by 2035, up from 142 billion m^3 yr^{-1} in 2010, an increase which would make HF extraction account for 49% of total natural gas production in the U.S. and make the U.S. a potential net exporter of natural gas by 2022 [7]. Globally, unconventional natural gas resources recovered through HF are expected to account for nearly half of newly developed gas production projects by the year 2035 [8].

In HF, the well casing is perforated at selected intervals within the targeted reservoir. Subsequent to perforating the well casing, high pressure is applied to the reservoir formation by the use of fluids, primarily water, to fracture the formation. The process is monitored at a control center, which is located at the well pad. Fracture propagation, including fracture length, is controlled based on experience working with the unique physical properties of the formation. The HF fluids are injected over a short period of time, usually within a week. Sand and other inert solids, such as ceramic beads, are injected into the formation to provide a support, or "proppant", which prevents the fractures from closing once the well pressure is released. In addition to proppant, other chemicals are added to the injected HF fluids. These chemicals are typically blended at the wellhead during, or immediately before, injection and serve various functions in the process, including preventing the growth of bacteria, facilitating the pumping of proppant down-hole and into the fractured formation, and minimizing mineral scaling of the well. Injected chemicals can include gelling and foaming agents, friction reducers, crosslinkers, breakers, pH adjusters, biocides, corrosion inhibitors, scale inhibitors, iron control chemicals, clay stabilizers, and surfactants [1], [9] and [10]. Many of the chemicals used in HF are added to keep the proppant in suspension and deliver the proppant into the artificial fractures [2] and [9]. There is a wide variety of chemicals that can be used in HF and the chemical compositions of individual mixtures are commonly held as trade secrets by HF practitioners. The reticence of HF contractors to disclose the proprietary constituents in the composition of their formulations has contributed to concerns amongst the general public about the potential environmental risks associated with the chemicals used in HF treatments [11].

This paper is focused on aqueous HF treatments; however, other fluids, including gases and petroleum distillates, can be used for unconventional oil and gas production. Gas fracturing employs dinitrogen or carbon dioxide gas and is typically used in shallow, water-sensitive formations that remain self-propped after fracturing, such as in areas of western Canada [12] and [13]. Petroleum distillates include hydrotreated light naphthenic distillates, mineral oil, diesel fuel, and kerosene [14]. Diesel fuel has been used in some areas and in some applications instead of water because higher concentrations of polymers can be injected using non-aqueous treatments [15] and [16]. Petroleum distillates may also be used as carrier fluids for dissolving additives before mixing, in what are otherwise aqueous HF treatments [1]. For example, crosslinkers and pH adjusters have been dissolved or suspended in hydrophobic carrier fluids before being mixed into aqueous fracturing fluids during well-injection in order to overcome the limitations of dry chemical blending and uncontrolled premature crosslinking [17]. The U.S. EPA has worked with major HF contractors and unconventional gas producers to eliminate the use of diesel fuel in fracturing fluid due to environmental and toxicity concerns [14] and [16].

Following HF, fluid returns to the surface as the pressure is released from the wellhead. This fluid is generally classified as either flowback or produced water [9]. Flowback is commonly defined as the return of injected fluids and produced water is water from the formation. The distinction between flowback and produced water during operations is not clear-cut, since mixing occurs in the formation. In practice, the term flowback is used to refer to initial, higher flows in the period immediately after well stimulation and produced water refers to long-term, typically lower flows associated with commercial hydrocarbon production. After the pressure in the well is reduced, flowback water is returned to the surface at high rates for up to several weeks, and this flow is, initially, predominantly fluids that were injected, but over time the fraction of the fluid that represents formation water increases [10], [18] and [19]. Produced water flows to the surface, along with the gas or oil, throughout the production life of the well and originates from water found in the geologic formation [10]. The volumetric recovery of injected water in the initial or flowback period varies widely and is strongly influenced by formation characteristics; while values as low as 5% and as high as 85% have been reported, recoveries between 30% and 50% appear representative [1],[5] and [10]. Although volumes of water used per well treatment for HF vary widely, values of approximately 7600 to 18,900 m^3 (2 to 5 million gallons) have been reported [1], [5], [10] and [20], suggesting between 1900 and 9000 m^3 (0.5 to 2.4 million gallons) of flowback will require proper management for reuse or disposal for each well. In addition to containing chemicals that were injected as part of the treatment, flowback water will also contain reaction products and constituents from the geologic formation which may include naturally occurring radioactive material (NORM); salts; heavy metals, such as mercury and lead; arsenic; and hydrocarbons, including polycyclic aromatic hydrocarbons and volatile and semi-volatile organic compounds [1] and [5]. Produced water is a longstanding and well-studied environmental management problem associated with both conventional and unconventional oil and gas production, and thus, was not considered in this paper, as it has been described elsewhere [21] and [22]. In this paper, we consider only the chemicals added to the injection well as part of unconventional production and evaluate those chemicals in the context of physical and chemical properties that affect their environmental fate, treatment potential, and toxicity. A systematic evaluation of HF chemicals in the context of environmental impact is needed [23], [24], [25],[26] and [27]. Recent use by HF contractors of the website FracFocus has increased the availability of information on chemicals used at well pads for HF. The most likely environmental exposure pathways for HF chemicals are expected to include the potential migration of contaminants from well pads to groundwater and surface water from accidental or operational spills, including transportation accidents and releases during treatment and disposal [10], [25], [28] and [29]. These are typical exposure pathways for industrial contaminants and in order to predict the environmental fate of HF chemicals, basic physical and chemical information needs to be compiled. Information needed for understanding environmental and health impacts of chemicals includes octanol–water partition coefficient, Henry's law constant, equilibrium constant, chemical oxygen demand (COD), melting point, boiling point, vapor pressure, aqueous solubility, biodegradability, and toxicity data for oral and inhalation exposure pathways [26]. These physical and chemical data are required for conducting environmental studies, environmental modeling, and for screening potential treatment strategies. These physical and chemical data are interpreted in the context of other variables such as soil, sediment or rock, temperature and moisture that ultimately govern the fate of chemical in the environment.

In response to concerns about the potential environmental and health impacts of HF, lists of HF chemicals have been collected over the past few years and made publicly available. Regulatory agencies in many states have established reporting requirements for unconventional production, but not all reporting requirements are mandatory [11], [30] and [31]. Many producers have been voluntarily publishing lists of HF chemicals on their company websites or in the FracFocus Chemical Disclosure Registry [32]. The FracFocus database contains records of chemicals used at wells located throughout the U.S. [32]. Participation in the FracFocus registry is voluntary; however, participation is required by 12 state regulatory agencies to meet chemical disclosure requirements in order to receive environmental permits [32]. Contained within FracFocus are lists of HF chemicals used for development of individual wells and the quantities of the chemicals used. Reporting is done by individual producers and the data input into the registry is not standardized in terms of the information provided. FracFocus does not include physical, chemical, and toxicological information.

The objectives of this study were to clarify what chemicals are used in HF applications, compile fundamental information on the chemicals used in HF, identify data gaps concerning what is known about HF chemicals, and interpret what information is known in the context of understanding the environmental fate of HF chemicals. In this paper, we developed a list of chemicals commonly used in HF and characterized these compounds in terms of their physical, chemical, and biological properties. Chemicals used in HF were organized by purpose and use in typical HF operations. Physical and chemical information needed for fate and transport studies, treatment technology assessments, waste management plans, risk assessments, and environmental modeling were compiled with results from standard mammalian oral and inhalation exposure tests. The compiled information was interpreted in the context of environmental releases and the challenges of developing technology for the treatment of flowback water. Hydraulic fracturing chemicals which are toxic, environmentally persistent, or for which critical physical and chemical information were not available, were identified as priority compounds for further investigation.

METHODS

Identification of Compounds Used in Hydraulic Fracturing

Eighty-one chemical additives used in HF fluid were identified and evaluated as part of this study (Table 1). Forty-one chemicals were identified using a list of commonly used chemicals provided by the FracFocus Chemical Disclosure Registry website [33]. This list was compared to a compiled on-line database by SkyTruth that contains data on 27,000 HF operations from January 2011 through

August 2012 [34]. A query was run using the SkyTruth database to verify the most commonly reported chemicals based on matching chemical names and CAS numbers. The SkyTruth database is largely based on information from the FracFocus registry and no reductions or additions to FracFocus list were needed as a result of the SkyTruth database check. The verified list was then compared to the U.S. EPA report that evaluated the potential impacts of HF on drinking water supplies [16]. After conducting an evaluation of the U.S. EPA report [16], a HF textbook [3], and published literature from industry sources [35], [36], [37], [38] and [39], our list of commonly used fracturing chemicals grew from 41 to 81.

Table 1: List of chemicals and chemical mixtures identified as being commonly used in hydraulic fracturing based on available sources

Chemical name	CAS number	Chemical formula	Source	Chemical name	CAS number	Chemical formula	Source
Acetaldehyde	75-07-0	C_2H_4O	[33]	Formic acid	64-18-6	CH_2O_2	[16] and [33]
Acetic acid	64-19-7	$C_2H_4O_2$	[33] and [16]	Fumaric acid	110-17-8	$C_4H_4O_4$	[16]
Acetone	67-64-1	C_3H_6O	[16]	Glutaraldehyde	111-30-8	C_5H_8O	[33]
Adipic acid	124-04-9	$C_6H_{10}O_4$	[16]	Glycol ethers	Various	Various	[16]
Alkyl dimethyl benzyl ammonium chloride	68424-85-1	Various	[35]	Guar gum	9000-30-0	Various	[16] and [33]
Ammonium chloride	12125-02-9	ClH_4N	[33]	Hemicellulase enzyme	9012-54-8		[39]
Ammonium persulfate	7727-54-0	$(NH_4)_2S_2O_8$	[33]	Hydrochloric acid	7647-01-0	HCl	[16] and [33]
Ammonium sulfate	7783-20-2	$(NH_4)2SO_4$	[16]	Hydrotreated light petroleum distillate	64742-47-8	Various	[33]
Borate salts	Various	Various	[33]	Hydroxyethyl cellulose	9004-62-0	Various	[16]

Boric acid sodium salt	1333-73-9	Na_3BO_3	[16] and [33]	Hydroxy-propyl cellulose	9004-64-2	Various	[3]
Calcium chloride	10043-52-4	$CaCl_2$	[33]	Hydroxy-propyl guar	39421-75-5	$(C_{27}H_{48}O_{18})n$	[16]
Calcium peroxide	1305-79-9	CaO_2	[3]	Isopropanol	67-63-0	C_3H_8O	[16] and [33]
Carbon dioxide	124-38-9	CO_2	[16]	Magnesium oxide	1309-48-4	MgO	[33]
Carboxy-methyl guar	39346-76-4	Various	[16]	Magnesium peroxide	14452-57-4	MgO_2	[33]
Carboxy-methyl hy-droxyethyl cellulose	9004-30-2	Various	[3]	Methanol	67-56-1	CH_4O	[16] and [33]
Carboxy-methyl hydroxypro-pyl guar	68130-15-4	Various	[16]	Monoetha-nolamine	141-43-5	C_2H_7NO	[16]
Choline chloride	67-48-1	$C_5H_{14}ClNO$	[33]	Monoethyl-amine	75-04-7	C_2H_7N	[16]
Citric acid	77-92-9	$C_6H_8O_7$	[33]	*N,n*-dimethyl formamide	68-12-2	C_3H_7NO	[37]
Copoly-mer of acrylamide and sodium acrylate	25987-30-8	Various	[33]	Naphtha-lene	91-20-3	$C_{10}H8$	[33]
Copper compounds	Various	Various	[16]	Nitrogen	7727-37-9	N_2	[16]
Didecyl dimethyl ammonium chloride	7173-51-5	$C_{22}H_{48}ClN$	[36]	Petroleum distillate	64741-85-1	Various	[33]
Diesel fuel	Various	Various	[16]	Phosphonic acid salt	Various	Various	[33]
Diethanol-amine	111-42-2	$C_4H_{11}NO_2$	[16]	Polyacryl-amide	9003-05-8	$(C_3H_5NO)n$	[33]
Dimethyl dihydroge-nated tallow ammonium chloride	Various	Various	[38]	Polyglycol ether	Various	Various	[16]
Potassium carbonate	584-08-7	K_2CO_3	[33]	Sodium persulfate	7775-27-1	$Na_2O_8S_2$	[3]
Potassium chloride	7447-40-7	KCl	[16]	Sodium polycarbox-ylate	Various	Various	[33]
Potassium hydroxide	1310-58-3	KOH	[16] and [33]	Sodium tetraborate decahydrate	1303-96-4	$B_4O_7 \cdot 2Na \cdot 10H_20$	[16] and [33]

Potassium metaborate	13709-94-9	BKO_2	[33]	Tetrakis hydroxy-methyl-phosphonium sulfate	55566-30-8	$(C_4H_{12}O_4P)_2O_4S$	[33]
Potassium persulfate	7727-21-1	$K_2O_8S_2$	[3]	Tetramethyl ammonium chloride	75-57-0	$C_4H_{12}ClN$	[33]
Propargyl alcohol	107-19-7	C_3H_4O	[16]	Thioglycolic acid	68-11-1	$C_2H_4O_2S$	[33]
Pyridinium	16969-45-2	C_5H_6N	[16]	Thiourea	62-56-6	CH_4N_2S	[16]
Quaternary ammonium chloride	61789-71-1	Various	[33]	Tributyl tetradecyl phosphonium chloride	81741-28-8	$C_{26}H_{56}PCl$	[39]
Sodium carbonate	497-19-8	Na_2CO_3	[33]	Triethanolamine zirconate	101033-44-7	$C_{24}H_{56}N_4O_{12}Zr$	[33]
Sodium chloride	7647-14-5	NaCl	[33]	Zirconium hydroxy lactate sodium complex	113184-20-6	$C_{12}H_{19}NaO_{16}Zr$	[33]
Sodium erythorbate	6381-77-7	$C_6H_7NaO_6$	[33]	Zirconium nitrate	13746-89-9	$Zr(NO_3)_4$	[16]
Sodium hydroxide	1310-73-2	NaOH	[33]	Zirconium sulfate	14644-61-2	$Zr(SO_4)_2$	[16]
Sodium lauryl sulfate	151-21-3	$C_{12}H_{25}NaO_4S$	[33]				
1-Bromo-3-chloro-5,5-dimethylhydantoin	16079-88-2	$C_5H_6BrClN_2O_2$	[35]	Ester salt	Various	Various	[16]
2,2-Dibromo-3-nitrilopropionamide	10222-01-2	$C_3H_2Br_2N_2O$	[16]	Ethanol	64-17-5	C_2H_6O	[16] and [33]
2-Bromo-3-nitrilopropionamide	1113-55-9	$C_3H_3BrN_2O$	[16]	Ethyl methyl derivatives	Various	Various	[16]
2-Butoxyethanol	111-76-2	$C_6H_{14}O_2$	[33] and [16]	Ethylene glycol	107-21-1	$C_2H_6O_2$	[16] and [33]

Physical and Chemical Characterization

Physical and chemical data for fracturing fluid additives was obtained from online chemical information databases [40], [41],

[42], [43], [44], [45] and [46], chemical reference books [47], [48], [49], [50] and [51], materials safety data sheets [39], [52] and [53], and a textbook [3]. In some cases, data was located in peer-reviewed scientific publications [28] and [54] and reports authored by the U.S. EPA [55] and [56], United Nations Environment Programme [57], and European Chemicals Agency [58]. Physical and chemical data are mostly based on laboratory tests using pure compounds and details of methods can be found in individual references.

Physical, chemical, and toxicological properties were selected for investigation based on their use in environmental fate and transport studies, treatability evaluations, remediation efforts, and risk assessments [59]. For example, the octanol–water partition coefficient is correlated with toxicity, soil sorption, and aquatic bioaccumulation [48]. The soil organic carbon partition coefficient (K_{OC}) is useful for determining soil sorption and contaminant transport rates [48]. Henry's constant (K_H) is an indicator of partitioning that occurs between aqueous and gaseous phases and is used in the design of air-stripping towers as well as many other design and modeling applications [48]. Chemical volatility from water was categorized as volatile, semi-volatile, or non-volatile according to K_H ranges of $x \geq 10^{-5}$, $10^{-5} > x \geq 3 \times 10^{-7}$, and $x < 3 \times 10^{-7}$ atm m^3 mol^{-1}, respectively [48]. Solubility is used in determining possible health risks. Theoretical COD was calculated using H_2O, CO_2, NO_3^-, SO_4^{2-}, and PO_4^{3-} as end products and is valuable for determining oxygen requirements for oxidizing treatments and remediation techniques. HF chemicals were categorized as non-biodegradable, inherently biodegradable, and readily biodegradable using OECD guidelines [60]. Biodegradability is useful for determining the effectiveness of biological treatment for wastewaters and the fate of chemicals released into the environment. Toxicity was described using median lethal dose (LD_{50}) for oral and inhalation exposure in rats, mice, and rabbits. Mammalian oral and inhalation toxicity data are commonly used in environmental and health risk assessments and mammalian oral toxicity data are collected under standard conditions [61]. To standardize interpretation, chemical toxicity was categorized according to United Nations standards in the *Globally Harmonized System of Classification and Labelling of Chemicals* (GHS), which classifies acute toxicity on a scale of 1 to 5, with 5 being the least toxic [61]. Where chemical concentrations were reported in the literature on a volumetric basis, the chemical density data collected as part of this study was used to calculate mass-based concentrations.

RESULTS AND DISCUSSION

For the HF chemicals identified (Table 1), toxicity data (Table 2 and Table 3) and chemical and physical data (Table 4, Table 5, Table 6, Table 7, Table 8 and Table 9) were summarized, where such data was available. Some chemicals and mixtures for which no information was found are discussed exclusively in the text. Properties which were not relevant for the chemical or chemical mixture are denoted by "–". Information which was not located in the standard databases is identified as "Not found" in the table, and information still missing after a search of the literature was conducted is marked as "Unknown". The tables of physical, chemical, and biochemical properties are interpreted in the context of evaluating chemicals for potential environmental hazard based on persistence, mobility, and other properties such as concentration and toxicity.

Table 2: Toxicity values via inhalation for frequently used additives in hydraulic fracturing fluids

Chemical	Inhalation toxicity rat (LC_{50})	Inhalation toxicity mouse (LC_{50})	Reference	Chemical	Inhalation toxicity rat (LC_{50})	Inhalation toxicity mouse (LC_{50})	Reference
2,2-Dibromo-3-nitrilopropionamide	320 mg m^{-3}/4 h	Not found	[40]	Methanol	87,500 mg m^{-3}/6 h, 64,000 ppm/4 h, >145,000 ppm/1 h	Not found	[40]

2-Butoxy-ethanol	450–486 ppm/4 h	700 ppm/7 h	[40] and [46]	Monoethyl amine	12,600 mg m^{-3}/4 h	Not found	[40]
Acetaldehyde	37,000 mg m^{-3}/30 min	1500 ppm/4 h	[40]	*N,n*-dimethyl formamide	Not found	9400 mg m^{-3}/2 h	[40]
Acetic acid	11,400 mg m^{-3}/4 h	5620 ppm/1 h	[40]	Naphthalene	>100 ppm/8 h	Not found	[41]
Acetone	76,000 mg m^{-3}/4 h, 50,100 mg m^{-3}/8 h	44,000 mg m^{-3}/4 h	[40],[41] and [46]	Potassium chloride	873 ppm/4 h	Not found	[50]
Adipic acid	>31,000 mg m^{-3}/1 h, 7700 mg m^{-3}/4 h	Not found	[41]	Propargyl alcohol	1040–1200 ppm/1 h, 873 ppm/2 h	2000 mg m^{-3}/2 h	[40],[46] and [50]
Diammonium peroxydisulphate	520,000 mg m^{-3}/1 h, >2950 mg m^{-3}/1 h	Not found	[40]	Sodium carbonate	2300 mg m^{-3}/2 h	1200 mg m^{-3}/2 h	[40]
Ethanol	20,000 ppm/10 h	39,000 mg m^{-3}/4 h	[40] and [46]	Sodium lauryl sulfate	>3900 mg m^{-3}/1 h	Not found	[46]
Formic acid	15,000 mg m^{-3}/15 min	6200 mg m^{-3}/15 min, 7400 mg m^{-3}/4 h	[46] and [40]	Sodium tetraborate decahydrate	>2 mg m^{-3}/1 h	Not found	[40]
Glutaraldehyde	280–800 mg m^{-3}/4 h; 5000 ppm/4 h	Not found	[40]	Tetrakis hydroxymethylphosphonium sulfate	5500 mg m^{-3}/4 h	Not found	[40]
Hydrochloric acid	3124 ppm/1 h, 4701 ppm/30 min	1108 ppm/1 h, 2644 ppm/30 min	[40]	Thioglycolic acid	210 mg m^{-3}/4 h	Not found	[46]
Isopropanol	51,045 mg m^{-3}/8 h; 72,600 mg m^{-3}/4 h	53,000 mg m^{-3}/2 h	[40]	Thiourea	> 900 mg m^{-3}/4 h	Not found	[41]

Table 3: Oral toxicity values for frequently used additives in hydraulic fracturing fluids

Chemical	**Oral toxicity (LD_{50}), rat (mg kg^{-1})**	**Oral toxicity (LD_{50}), mouse (mg kg^{-1})**	**Oral toxicity (LD_{50}), rabbit (mg kg^{-1})**	**Reference**	**Chemical**	**Oral toxicity (LD_{50}), rat (mg kg^{-1})**	**Oral toxicity (LD_{50}), mouse (mg kg^{-1})**	**Oral toxicity (LD_{50}), rabbit (mg kg^{-1})**	**Reference**
Acetaldehyde	661–1930	1230	Not found	[40]	Ethylene glycol	4700	7500	Not found	[40] and [50]
Acetic acid	3310–3530	4960	1200	[40]	Formic acid	1100	700	Not found	[40] and [50]
Acetone	5800–9800	3000–5200	5340	[40]	Fumaric acid	9300–10,700	Not found	Not found	[50]
Adipic acid	>11,000	1900	>11,000	[44] and [50]	Glutaraldehyde	134–1470	100	1.59 mL kg^{-1} 50% aqueous solution	[40] and [50]
Ammonium chloride	1650	1300	LDL_o = 1000	[50]	Guar gum	6770	8100	7000	[50]
Ammonium sulfate	3000	3040	Not found	[40] and [50]	Hydrochloric acid	238–277	Not found	900	[40]
Ammonium persulfate	689	Not found	Not found		Hydroxypropyl cellulose	10,200	>5000	Not found	[46]
Boric acid sodium salt	Not found	Not found	Not found		Hydroxypropyl guar	Not found	Not found	Not found	

Calcium chloride	1000–4179	1940–2045	100–1000	[40] and [50]	Isopropanol	4710–5840	3600–4475	5030–7990	[40] and [50]
Calcium peroxide	Not found	Not found	Not found		Magnesium oxide	3870–3990	810	Not found	[40]
Carboxymethyl guar	17,800	Not found	Not found	[46]	Magnesium peroxide	Not found	Not found	Not found	
Carboxymethyl hydroxyethyl cellulose	Not found	Not found	Not found		Methanol	5628–6970	7300	14400	[40] and [44]
Carboxymethyl hydroxypropyl guar	Not found	Not found	Not found		Monoethanolamine	1720–10,200	700–1475	1000	[41] and [50]
Choline chloride	3400–6640	3900	Not found	[40]	Monoethylamine	400–530	Not found	Not found	[40] and [50]
Citric acid	3000–6730	5040	7000	[50]	*N,n*-dimethyl formamide	2800–3000	3750	>5000	[40]
Copolymer of acrylamide and sodium acrylate	Not found	Not found	Not found		Naphthalene	490–2600	350–710	Not found	[40]
Diammonium peroxydisulphate	495–820	Not found	Not found	[40]	Sodium persulfate	Not found	Not found	Not found	
Nitrogen	Not found	Not found	Not found		Potassium carbonate	1870	2570	Not found	[50]
Petroleum distillate	Not found	Not found	Not found		Sodium polycarboxylate	Not found	Not found	Not found	
Phosphonic acid salt	Not found	Not found	Not found		Sodium lauryl sulfate	1288	Not found	Not found	[50]

Polyacrylamide	>1000	12,950	11,250	[44] and [50]	Sodium tetraborate decahydrate	5660	2000	Not found	[40]
Potassium chloride	2600	383–1500	Not found	[40], [46] and [50]	Tetrakis hydroxymethyl-phosphonium sulfate	248–333	Not found	Not found	[40] and [50]
Potassium hydroxide	273–1230	Not found	Not found	[44] and [50]	Tetramethyl ammonium chloride	50	125	Not found	[46] and [50]
Potassium metaborate	Not found	Not found	Not found	[40]	Thioglycolic acid	114	242	119	[40] and [50]
Potassium persulfate	802	Not found	Not found	[50]	Thiourea	20–640	8500	10,000	[40] and [41]
Propargyl alcohol	20–110	50	Not found	[40], [44] and [50]	Tributyl tetradecyl phosphonium chloride (48–50%)	1002	Not found	Not found	[53]
Pyridinium	Not found	Not found	Not found		Triethanolamine zirconate	Not found	Not found	Not found	
Sodium carbonate	2800–4090	6600	Not found	[40], [46] and [50]	Zirconium hydroxy lactate sodium complex	Not found	Not found	Not found	
Sodium chloride	3000	4000	Not found	[40]	Zirconium nitrate	2290	Not found	Not found	[50]
Sodium erythorbate	>5000	Not found	Not found	[44]	Zirconium sulfate	3500	Not found	Not found	[50]
Sodium hydroxide	140–340	Not found	500	[40] and [50]					

1-Bromo-3-chloro-5,5-dimethyl-hydantoin	578–1390	Not found	>2000	[40],[44] and [52]	Didecyl dimethyl ammonium chloride	84–331	268	Not found	[40],[44] and [50]
2,2-Dibromo-3-ditrilopropion-amide	178–235	Not found	118	[40] and [50]	Diethanolamine	710–1820	3300	2200	[44] and [50]
2-Bromo-3-ni-trilopropion-amide	Not found	Not found	Not found		Dimethyl dihy-drogenated tal-low ammonium chloride	Not found	Not found	Not found	
2-Butoxyethanol	470–3000	1200–1519	300–320	[40] and [44]	Ethanol	7060–10,600	3450	Not found	[40] and [50]

Table 4: Properties of gelling agents, foaming agents, and associated chemicals used in hydraulic fracturing

Chemical	log (K_{OW})	K_H(atm m^3 mol^{-1})	K_{OC}	pK_a	COD (gO_2 g^{-1})	Density (g cm^{-3}at 25 °C)	Melting point (°C)	Boiling point (°C)	Vapor pressure (mmHg at 25 °C)	Solubility in water (mg L^{-1})	Biodegradability	Reference
Adipic acid[a]	0.08	4.7×10^{-12}	26	4.44	1.42	1.36	152.5	337.5	3.18×10^{-7}	3.00×10^4at 30 °C	RB	[40],[42] and [47]
Fumaric acid[a]	0.46	8.5×10^{-14}	33.5	pKa_1 = 3.02, pKa_2 = 4.46	0.83	1.635	287	522	1.54×10^{-4}	7×10^{-3}	RB	[40], [41],[42] and [50]
Guar gum[a]	Not found	Not found	Not found	–	1.18	Not found	Not found	Not found	Not found	Not found	RB	[70] and [71]
Carboxymethyl guar[a]	Not found	Not found	Not found	–	Not found	Not found	Not found	Not found	Not found	Not found	RB	[70] and [71]
Hydroxypropyl guar[a]	Not found	Not found	Not found	–	1.45	Not found	Not found	Not found	Not found	Not found	RB	[70] and [71]
Carboxymethyl hydroxypropyl guar[a]	Not found	Not found	Not found	–	Not found	Not found	Not found	Not found	Not found	Not found	RB	[70] and [71]
Hydroxyethyl cellulose[a]	Not found	Not found	Not found	–	Not found	Not found	Not found	Not found	Not found	Not found	Not found	–

Hydroxypropyl cellulose[a]	Not found	Not found	Not found	–	Not found	Not found	Not found	Not found	Not found	Not found	Not found	–
Carboxymethyl hydroxyethyl cellulose[a]	Not found	Not found	Not found	–	Not found	Not found	Not found	Not found	Not found	Not found	Not found	–
2-Butoxyethanol[b]	0.45	2.36×10^{-6}	8	–	2.30	0.90	−74.8	168.4	0.88	Miscible	RB	[40],[48] and [50]
Diethanolamine[b]	−1.43	3.9×10^{-11}	4	8.96	2.13	1.10	28	268.8	1.4×10^{-4}	Miscible	RB	[40],[41] and [46]
Ethanol[c]	−0.31	5.00×10^{-6}	2.75	15.9	2.08	0.82	−114.1	78.29	59.3	Miscible	RB	[40], [41],[46],[49] and [51]
Isopropanol[c]	0.05	8.10×10^{-6}	25	17.1	2.40	0.79	−87.9	82.3	45.4	Miscible	RB	[41], [42],[46],[49] and [50]
Methanol[c]	−0.73	4.66×10^{-6}	2.75	15.3	1.50	0.79	−97.8	64.7	127	Miscible	RB	[40], [41],[46] and [48]
Ethylene glycol[c]	−1.36	6.00×10^{-8}	0.2	15.1	1.29	1.11	−12.7	197.3	0.092	Miscible	RB	[40], [41],[46] and [49]

COD = Theoretical chemical oxygen demand, RB = Readily biodegradable.

[a]Gelling agent.

[b]Foaming agent.

[c]Stabilizing agent used with gelling and foaming agents.

Table 5: Properties of crosslinkers used in hydraulic fracturing

Chemical[a]	log (K_{OW})	K_H (atm m^3 mol^{-1})	K_{OC}	pK_a	COD (gO_2 g^{-1})	Density (g cm^{-3} at 25 °C)	Melting point (°C)	Boiling point (°C)	Vapor pressure (mmHg at 25 °C)	Solubility in water (mg L^{-1})	Biodegradability	Reference
Ammonium chloride	−4.37	3.88×10^{-13}	–	–	1.20	1.52	338	520	2.42×10^{-3}	3.95×10^5	RB	[41] and [42]
Boric acid sodium salt	–	–	–	–	–	Not found	Not found	Not found	Not found	Not found	NB	
Potassium metaborate	–	–	–	–	–	2.3	947	Not found	Not found	Not found	NB	
Sodium tetraborate decahydrate	−1.53	–	–	–	–	1.73	75	Not found	Not found	5.93×10^4	NB	[40],[50] and [58]
Zirconium hydroxy lactate sodium complex	Unknown	Unknown	Unknown	–	Unknown	Unknown	Unknown	Unknown	Unknown	Unknown	Unknown	
Zirconium nitrate	–	–	–	–	–	Not found	100 (decomposes)	–	Not found	Not found	Not found	
Zirconium sulfate	–	–	–	–	–	3.22	410 (decomposes)	–	Not found	5.25×10^5	Not found	

Monoethanolamine	−1.31	3.25×10^{-8}	5	9.50	2.36	1.02	10.5	171	0.40	Miscible	RB[a]	[40],[41] and [42]
Monoethylamine	−0.13	1.23×10^{-5}	20	10.87	3.55	0.7	−80.5	16.5	1048	1×10^{6}	RB	[40],[41] and [42]
Triethanolamine zirconate	Unknown	Unknown	Unknown	Unknown	1.78	Unknown	Unknown	Unknown	Unknown	Unknown	Unknown	

COD = Theoretical chemical oxygen demand, RB = Readily biodegradable, NB = Non-biodegradable.

[a]Methanol and ethylene glycol are stabilizers or winterizing agents used in conjunction with crosslinkers, and are also used with gelling and foaming agents. They are described in Table 4.

Table 6: Properties of breakers used in hydraulic fracturing

Chemical	**COD (gO_2 g^{-1})**	**Density (g cm^{-3} at 25 °C)**	**Solubility in water (mg L^{-1})**	**Reference**
Ammonium sulfate	0.97	1.77	7.67×10^5	[40], [41] and [43]
Ammonium persulfate	0.49	1.98	8.35×10^5	[40]
Calcium chloride	–	2.15	8.13×10^5	[40]
Calcium peroxide	–	3.34	Insoluble	[45] and [50]
Diammonium peroxydisulphate	0.49	1.98	8.35×10^5	[40]
Magnesium oxide	–	3.65–3.75	86	[40] and [50]
Magnesium peroxide	–	3	Insoluble	[40]
Potassium persulfate	–	2.48	5.2×10^4	[40] and [45]
Sodium chloride	–	2.16	3.57×10^5 at 0 °C	[40], [41] and [42]
Sodium persulfate	–	2.4	5.49×10^3	[43]
Glycol ethers	Not Found	Not Found	Not Found	
Hemicellulase enzyme	Not Found	Not Found	Not Found	

COD = Theoretical chemical oxygen demand.

Table 7: Properties of pH adjusters, corrosion inhibitors, and iron controlling agents used in hydraulic fracturing

Chemical[a]	log (K_{OW})	K_H(atm m^3 mol^{-1})	K_{OC}	pK_a	COD (gO_2 g^{-1})	Density (g cm^{-3}at 25 °C)	Melting point (°C)	Boiling point (°C)	Vapor pressure (mmHg at 25 °C)	Solubility in Water (mg L^{-1})	Biodegradability	Reference
Acetaldehyde[b]	–0.34	6.67×10^{-5}	16	13.57	1.82	0.78	–123.4	20.1	902	1×10^{6}	RB	[40], [41],[46] and [49]
Acetic acid[c,d]	–0.34	1.23×10^{-3}	6.5–228	4.76	1.07	1.05	16.6	117.9	15.7	Miscible	RB	[40],[48] and [51]
Acetone[b]	–0.24	4.26×10^{-5}	18	20	2.20	0.78	–94.7	56.1	232	Miscible	RB	[40], [42],[46] and [47]
Citric acid[d]	–1.72	8.33×10^{-18}	3.1	2.79	0.75	1.67	153	Decomposes	1.7×10^{-8}	3.83×10^{5}	RB	[40] and [49]
Ethyl methyl derivatives[b]	Various	Various	Various	–	Various	Various	Various	Various	Various	Various	Various	
Formic acid[b]	–0.54	1.67×10^{-7}	1	3.75	0.35	1.22	8.3	101	42.59	Miscible	RB	[40.41,49]
Hydrochloric acid[c]	0.25	4.9×10^{-10}	1.08	<1	–	1.18	–114.22	–85.05	35,424	6.73 × 105	–	[40] and [41]

N,n-dimethyl for-mamide [b]	−1.01	7.4×10^{-8}	7	0.3	2.41	0.94	−60.4	153	3.87	Mis-cible	RB	[40], [41],[42],[46] and [48]
Potassium carbonate[c]	−6.19	Not Found	Not Found	–	–	2.29	899	Decom-poses	2.44×10^{-17}	1.11×10^{6}	–	[40] and [42]
Potassium hydroxide[c]	Not Found	Not Found	Not Found	–	–	2.04	380	1327	1 at 714 °C	1.12×10^{6}	–	[40],[41] and [50]
Propargyl alcohol[c]	−0.38	1.2×10^{-6}	14	–	2.00	0.97	−48to −52	114–115	15.6	Mis-cible	RB	[40],[41] and [42]
Pyridini-um[b]	Not found	Not found	Not found	–	3.20	Not found	Not found	Not found	Not found	Not found	Not found	
Sodium carbonate[c]	−6.19	6.1×10^{-9}	Not Found	–	–	2.54	856	Decom-poses	9.92×10^{-17}	3.07×10^{5}	–	[40] and [42]
Sodium erythor-bate[d]	−7.05	6.79×10^{-14}	Not found	–	0.80	Not found	168–175	Not found	2.33×10^{-18}	1.60×10^{5}	Not found	[40], [42],[43] and [49]
Sodium hydroxide[c]	−3.88	8.45×10^{-9}	0.22	–	–	2.13	323	1388	1.82×10^{-21}	1.11×10^{6}	–	[40] and [42]
Thiogly-colic acid[d]	0.09	1.94×10^{-8}	27	3.55	1.22	1.33	−16.5	120	8.68×10^{-2}	Mis-cible	IB	[40],[41] and [49]
Thiourea[b]	−1.08	2.0×10^{-9}	3–6	2.03	2.52	1.41	176–178	Sublimes	2.8×10^{-3}	1.42×10^{5}	IB	[40], [41],[42] and [46]

COD = Theoretical chemical oxygen demand, RB = Readily biodegradable, IB = Inherently biodegradable.

[a]Fumaric acid is a pH adjuster and also a gelling agent. It is described in Table 4. Isopropanol and

methanol are stabilizers or winterizing agents used in conjunction corrosion inhibitors and also used with gelling and foaming agents. They are described in Table 4.

[b]Corrosion inhibitor.

[c]pH adjuster.

dIron controlling agent.

Table 8: Properties of biocides used in hydraulic fracturing

Chemical[a]	**log (K_{OW})**	**K_H(atm m^3 mol^{-1})**	**K_{OC}**	**COD (gO_2 g^{-1})**	**Density (g cm^{-3}at 25 °C)**	**Melting point (°C)**	**Boiling point (°C)**	**Vapor pressure (mmHg at 25 °C)**	**Solubility in water (mg L^{-1})**	**Biodegradability**	**Reference**
1-Bromo-3-chloro-5,5-dimethylhydantoin	−0.94	8.10×10^{-7}	23	0.99	Not found	163–164	Not found	6.6×10^{-6}	8.26×10^3	Not found	[40]
2,2-Dibromo-3-nitrilopropionamide	0.8	1.91×10^{-8}	65	0.66	Not found	123–126	190 (decomposes)	9×10^{-4}	15,000	B	[40] and [42]
2-Bromo-3-nitrilopropionamide	Unknown	Unknown	Unknown	1.08	Not found	Not found	Not found	Not found	Not found	Not found	

Alkyl dimethyl benzyl ammonium chloride	Various	Various	6.4×10^5–6.2×10^6	Various	0.94	Various	241	3.53×10^{-12}	Various	RB	[55]
Didecyl dimethyl ammonium chloride	4.4×10^5	6.85×10^{-10}	4.4×10^5–1.6×10^6	3.09	0.92	228.81	Not found	2.33×10^{-11}	0.55	B	[40] and [42]
Glutaraldehyde	−0.18	1.10×10^{-7}	120–500	2.08	1.06	−14	188	0.6 at 30° C	2.2×10^5	RB	[40],[41] and [49]
Tetrakis hydroxymethylphosphonium sulfate	−20.39	1.7×10^{-23}	140	0.95	1.41	−35	111	1.27×10^{-5}	1×10^6	RB	[40],[42] and [56]
Tributyl tetradecyl phosphonium chloride	Not found	Not found	Not found	3.02	Not found	60	Not found	Not found	Not found	Not found	[54]

COD = Theoretical chemical oxygen demand, RB = Readily biodegradable, B = Biodegradable.

[a]Ammonium chloride is a biocide and also a crosslinker. It is described in Table 5.

Table 9: Properties of clay stabilizers used in hydraulic fracturing

Chemical[a]	**log (K_{OW})**	**K_H(atm m^3 mol^{-1})**	**K_{OC}**	**COD (gO_2 g^{-1})**	**Density (g cm^{-3}at 25 °C)**	**Melting point (°C)**	**Boiling point (°C)**	**Vapor pressure (mmHg at 25 °C)**	**Solubility in water (mg L^{-1})**	**Biodegradability**	**Reference**
Potassium chloride[b]	−0.46	–	–	–	1.99	771	1420	Not found	3.55×10^5	–	[40] and [43]
Choline chloride[b]	−5.16	2.03×10^{-16}	2.3	2.06	1.1	305	Decomposes	4.93×10^{-10}	$>6.5 \times 10^5$	RB	[40], [41],[42] and [57]
Tetramethyl ammonium chloride[b]	−4.18	4.2×10^{-12}	8	2.34	1.17	402	Decomposes	1.2×10^{-8}	5.9×10^5	NB	[40]
Dimethyl dihydrogenated tallow ammonium chloride[c]	Various	Various	Various	Various	Various	Various	Various	Various	Various	Various	
Sodium lauryl sulfate[c]	1.6	Not found	10,000	2.00	Not found	206	Not found	4.7×10^{-13}	1×10^5	RB	[40], [41],[43] and [49]

COD = Theoretical chemical oxygen demand, RB = Readily biodegradable, NB = Non-biodegradable.

[a]Sodium chloride is a clay stabilizer and also a breaker. It is described in Table 6. 2-Butoxyethanol, ethanol, and isopropanol are stabilizers or winterizing agents used in conjunction with surfactants and are also used with gelling and foaming agents. They are described in Table 4.

[b]Clay stabilizer.

[c]Surfactant.

Gelling and Foaming Components

Gelling agents are used in HF to increase fracturing fluid viscosity, allowing for better proppant suspension and transport into developed fractures. Gels can be linear, consisting of single stranded polymers, or crosslinked where individual polymer strands are chemically joined at specific functional groups to create larger molecules, further increasing fluid viscosity. Gelling agents are selected based on site-specific conditions in the well, including temperature and salinity [3]. Gelling compounds can be added to the fracturing fluid as powders or concentrates after being dissolved into a non-aqueous solvent [16]. In deep, high temperature reservoirs (above 132 °C), gel stabilizers are added to prevent premature decomposition[62]. When gel fracturing treatments are used, the concentration of the gelling agent ranges from approximately 10–1000 mg L^{-1}[1], [5], [10] and [63].

Commonly used gelling agents include guar, derivatives of guar [carboxymethyl guar (CMG), hydroxypropyl guar (HPG), and carboxymethyl hydroxypropyl guar (CMHPG)], and cellulosic compounds [hydroxyethyl cellulose (HEC), hydroxypropyl cellulose (HPC) and carboxymethyl hydroxyethyl cellulose (CMHEC)] [3], [16] and [64]. Guar molecules are composed of β-1,4-linked mannose units with α-1,6-linked galactose units attached (Fig. 1a) [65], [66] and [67], where the ratio of mannose to galactose units is typically 1.6:1 to 1.8:1 [66] and [67]. The cellulose derivatives HEC, HPC, and CMHEC contain a glucose sugar backbone (Fig. 1b). Guar derivatives are produced by modifying the guar hydroxyl groups to carboxymethyl and hydroxypropyl groups ($O—CH_2—COO$) (Fig. 1c and d) [68] and [69].

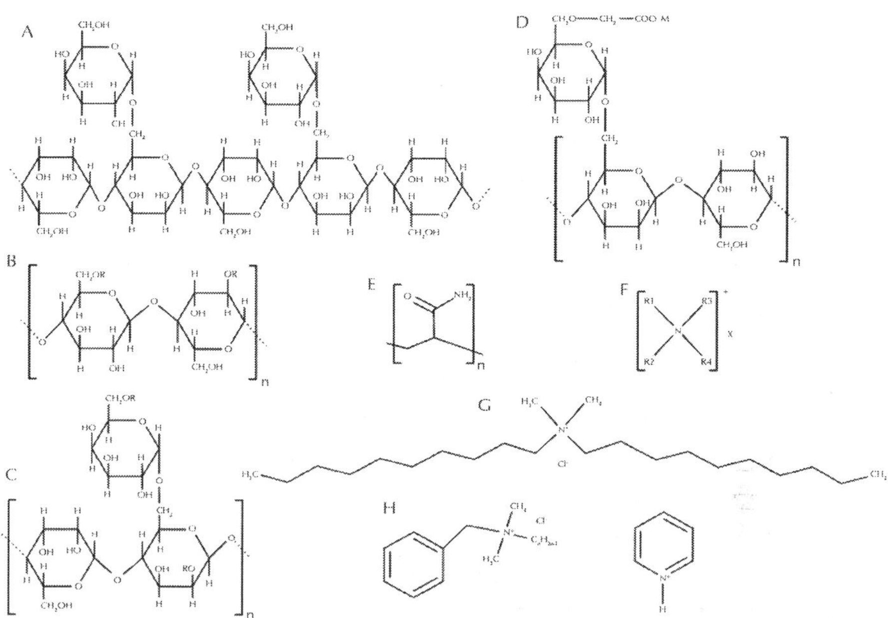

Figure 1: Structures of commonly used chemicals in hydraulic fracturing fluids. (a) Guar, a gelling agent. The backbone has α-1,6-lined galactose side chain units attached randomly (adapted from Cheng and Prud'homme [67]); (b) hydroxyethylcellulose (HEC), a gelling agent. R = CH_2CH_2OH (adapted from Economides and Nolte [3]); (c) hydroxypropyl guar (HPG), a gelling agent. R = CH_2—CHOH—CH_3 (adapted from Economides and Nolte [3]); (d) carboxymethyl guar gum (CMG), a gelling agent (adapted from Adhikary et al. [69]); (e) polyacrylamide, a friction reducer; (f) generic quaternary ammonium compound, a biocide (adapted from Fredell [87]); (g) didecyl dimethyl ammonium chloride (DDAC), a biocide; (h) Alkyl dimethyl benzyl ammonium chloride (ADBAC), a biocide; (i) pyridinium, a corrosion inhibitor.

Guar and guar derivatives are commonly used as food additives and in cosmetic and health products. Guar, guar derivatives, and cellulosic compounds are derived from natural materials and are non-toxic. Guar and its derivatives are readily biodegradable; however, cellulose polymers may be more resistant to biodegradation [70] and [71]. Gelling agents will contribute significantly to the COD of flowback water. The COD of guar gum and HPG [($C_{27}H_{48}O_{18}$)

n] are 1.18 and 1.45 gO_2 g^{-1}, respectively, and they are used in high concentrations. The high concentrations of gelling agents may make implementation of some treatment processes, such as reverse osmosis, difficult since the organic matter will encourage microbial processes that cause fouling on membrane surfaces [72] and [73]. Based on criteria of persistence and toxicity, these gelling agents are not of concern, but they need to be considered in treatment design and as a source of oxygen demand during an environmental release.

In addition to guar and cellulosic derivatives, adipic acid and fumaric acid are also used as gelling agents[16]. These acids are used as thickeners in foods and are above GHS Category 5 and therefore non-toxic by our interpretation; however, one source reported adipic acid as belonging in GHS Category 4 for mice (Table 3) [50]. Like all acids, these compounds may be irritating to the eyes, nose, and throat when in powdered form [40]. Chemical and physical data indicate low volatility potential from water and high sorption potential (Table 4), suggesting low mobility in soil and groundwater. Both acids are readily biodegradable, contribute significant COD, and are used in high concentrations in HF fluids, which, as described above, can influence treatment train design and pose an environmental risk when released.

In addition to gel treatments, foamed fracturing fluids, which are created by injecting dinitrogen or carbon dioxide gas into gelled liquids, are also used to alter fracturing fluid viscosity. Foamed fluids reduce the amount of liquid required for fracturing; for example, reductions of 75% are possible when compared with linear and crosslinked gel treatments [15] and [16]. In low pressure reservoirs, use of foamed fracturing fluids accelerates fluid recovery, and in formations with large amounts of clay, foamed fluids reduce clay swelling [3].

Diethanolamine, 2-butoxyethanol, and related ester salts are used in foamed treatments. Diethanolamine and 2-butoxyethanol have low log K_{OW} values, suggesting they will not sorb to soils and will likely be transported with groundwater flows. 2-Butoxyethanol is semi-volatile from water, based on its Henry's constant (Table

4), and meets criteria for GHS Category 2 toxins based on rat inhalation toxicity LC_{50}values (Table 2) [61]. Both 2-butoxyethanol and diethanolamine are GHS Category 4 chemicals based on rat oral toxicity LD_{50} values [61]. These compounds are readily biodegradable (Table 4), but information on the concentrations at which these additives are used in HF was not found. Diethanolamine and 2-butoxyethanol are not expected to be persistent in the environment, but based on expected mobility, possible toxicity, and lack of available information regarding how much of these materials are used, these compounds deserve further investigation.

Gel stabilizers used to prevent decomposition include ethanol, isopropanol, methanol, ethylene glycol, and polyglycol ether [16] and [32]. Ethanol, isopropanol, and methanol are used extensively in industry, beverages (ethanol), and household cleaners, and based on the Henry's constant values, all three chemicals are expected to be semi-volatile from water and are readily biodegradable (Table 4). These compounds sorb poorly to soils and are transported with groundwater. The environmental fates of ethanol and methanol have been extensively studied, due to their use as gasoline additives. Undiluted ethanol, isopropanol, and methanol are toxic by ingestion to rats between doses of 4710 and 7060 mg kg^{-1}(Table 3). These compounds used as product stabilizers are GHS Category 5 or non-toxic, are widely used in industry, are easily biodegradable, and are well characterized, and are therefore not considered high priority compounds for further investigation.

Friction Reducers

An alternative method to gel and foam treatments is "slickwater" fracturing. In slickwater treatments, chemicals are added to reduce fluid surface tension and facilitate removal of fracturing fluid from the formation. Slickwater treatments have a reputation for improving well production by reducing fracture plugging compared to gels [62]. In slickwater fracturing, up to 3785 m^3 (1 million gallons) of water may be required for each stage of production, with six to nine stages, or repeated injections, occurring per well [15]. When friction reducers are added, development of higher pressure within the formation is possible because the fluid viscosity and headloss due to friction are reduced, increasing effective static pressure[15].

A commonly used friction reducer is 2-propenamide (polyacrylamide, $(C_3H_5NO)n$, (Fig. 1e) [32]. Reported application rates in fracturing fluids are 30–1200 mg L^{-1}[3], [5], [10], [15], [16] and [63]. There are three forms of polyacrylamide: anionic, cationic, and nonionic [15]. According to King [10], nearly all slickwater fracturing projects use polyacrylamide as a friction reducer.

While log K_{OW}, K_H, and K_{OC}, and physical information, such as density, melting point, vapor pressure, and solubility was not located for polyacrylamide, it is known that it is water soluble, non-volatile, and above GHS Catgeory 5, and therefore, non-toxic (Table 3). It has been suggested that polyacrylamide can degrade through heating or exposure to ultraviolet radiation into acrylamide, which is a known human carcinogen, mutagen, and teratogen [74] and [75]. Polyacrylamide is biodegradable with a COD of 2.25 $gO_2\ g^{-1}$. It has been shown that no acrylamide is produced during the aerobic biodegradation of polyacrylamide [75] and [76]. Based on these results, a flowback water from slickwater treatments can be expected to have a high COD, and therefore present treatment challenges, but this friction reducer is not considered a priority material for investigation based on persistance or toxicity.

Crosslinkers

Crosslinkers are added to fracturing fluids to chemically bind individual gel polymer molecules together to form larger molecules (Fig. 2), resulting in higher viscosity, more elasticity, and better proppant transport compared with linear gels that are not crosslinked. Crosslinkers frequently used in HF include borate salts; aluminum, titanium, and zirconium compounds; monoethanolamine; and monoethylamine [3],[16] and [32]. Ammonium chloride, ethylene glycol, and potassium hydroxide are also used as crosslinkers[16]. Concentrations of crosslinkers in fracturing fluid are relatively low and range from 0.5 to 250 mg L^{-1}[1],[5], [10], [16] and [63].

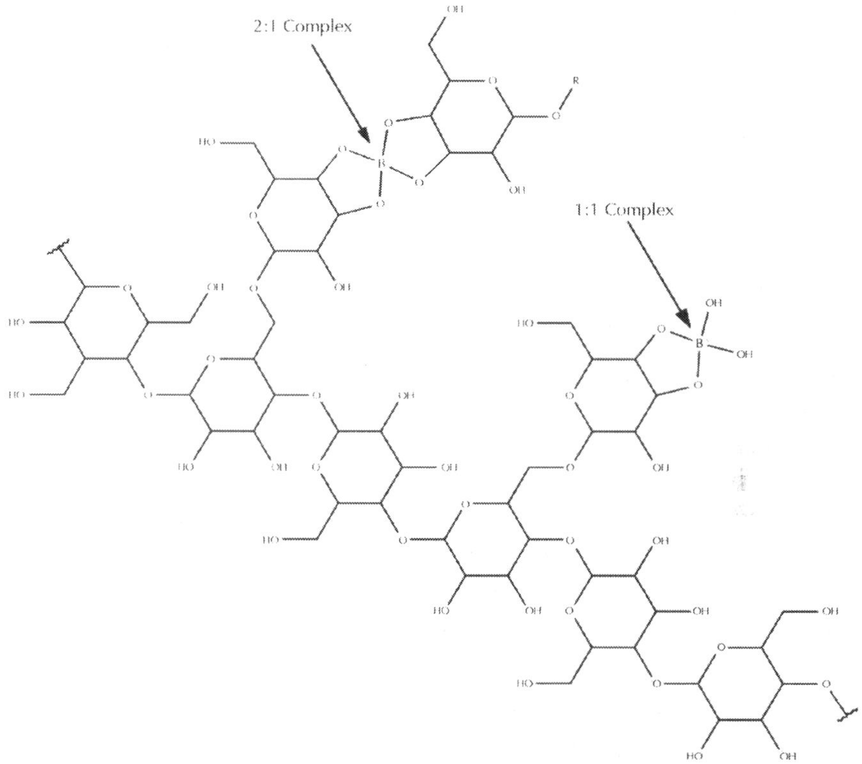

Figure 2: Structure of guar crosslinked with borate. The R group denotes another molecule of guar (adapted from Barati et al. [80]).

Available chemical and physical information for borate (Fig. 2) and zirconium compounds is limited (Table 5). These chemicals are not expected to significantly contribute to HF fluid COD. Inhalation and topical exposure to boron compounds is generally associated with upper respiratory tract, skin, and eye irritation[40]. Ingestion of 15 to 20 g of boric acid has resulted in severe toxicity or death in adults [40]. The borate salt sodium tetraborate decahydrate is a GHS Category 4 compound for mice, but is non-toxic (above GHS Category 5) for rats (Table 3). Zirconium compounds have been used to replace toxic chromium crosslinking compounds in HF treatments [77]. Zirconium is not considered a biologically reactive metal and is not expected to be a concern toxicologically

[78]. Oral toxicity of zirconium nitrate and zirconium sulfate places these HF chemicals in GHS Category 5 (Table 3), although data on the toxicity of zirconium lactate complexes and triethanolamine zirconate was not found. Based on their use at low concentrations and their moderate toxicity, these inorganic crosslinkers are not expected to be a major environmental concern, however environmental and toxicity data on the lactate and triethanolamine complexes are needed.

Ammonium chloride is water soluble and has a low Henry's constant at neutral pH (Table 5), indicating that the potential for volatilization from fracturing fluid is low. Ammonia is readily degradable to nitrate by bacteria and has a significant COD (Table 5), suggesting this crosslinker could contribute to the COD of flowback water. Toxicity values for ammonium chloride are in GHS Category 4 (Table 3). However, ammonium chloride is seemingly used in low concentrations, is common in the environment, and has been thoroughly studied as an environmental and health hazard, so is not considered a priority HF chemical in this context.

Monoethanolamine and monoethylamine are both readily biodegradable and soluble in water. The Henry's constant for monoethanolamine implies that this chemical will not volatilize readily from solution, but the significantly higher Henry's constant of monoethylamine indicates a greater potential for volatilization (Table 5); however, inhalation toxicity for rats is moderate (Table 2). Monoethylamine has reported LD_{50}oral toxicity values for rats between 400 and 530 mg kg^{-1} (Table 3), placing this HF chemical in GHS Category 4. The range of reported LD_{50} values for monoethanolamine is wide (Table 3), which suggests this HF chemical falls in GHS Category 4 as well. Of the crosslinkers surveyed in this study, monoethanolamine and monoethylamine appear to be the compounds of most interest and both are expected to be highly mobile in soils and groundwater based on K_{OC} values; however, neither compound is expected to be environmentally persistent.

Breakers

Following fracturing, an enzyme or inorganic breaker is introduced to reverse crosslinking, which reduces viscosity of gelled fluids and allows removal of residual polymers from newly created fractures, which would otherwise impede flow of gas and reduce well productivity [62] and [79]. The breaker reacts with and disrupts polymers, resulting in reduced molecular weight and fluid viscosity, allowing fracturing fluid to be recovered from wells [4]. Encapsulated or time-release breakers may be included in gels to reduce the chances of the polymer breaking prematurely [3], [80] and [81]. Breakers are also required for slickwater fracturing fluids to degrade friction-reducing polymers. At temperatures below 66 °C, enzyme breakers can be used to break guar gels [66] and [82] but under higher temperature conditions (94–121 °C) and at pH above 10.5, the effectiveness of most enzymes is reduced, and inorganic breakers are used [66]. Breakers are typically used in concentrations ranging from 1 to 400 mg L^{-1} [5] and [63].

Enzymes and inorganic compounds can be applied as breakers either while the fluid is being pumped into wells or after fracturing is complete. The commonly used enzyme breakers are hemicellulases; these non-toxic chemicals could contribute to COD of flowback water since they are proteins and degrade cellulose polymers into smaller, more rapidly degraded sugars. Many inorganic breakers are simple salts, such as calcium chloride, sodium chloride, and ammonium sulfate [16] and [32], which present no specific health hazard, but may contribute to salinity and, in the case of ammonium, COD of flowback water.

Inorganic oxidants, including ammonium, potassium, and sodium salts of peroxydisulfate [83], calcium and magnesium peroxide [3] and [32], and magnesium oxide [32] are used as breakers. Inorganic oxidants are well-characterized (Table 6) and present a number of chemical and physical hazards. These HF chemicals are not volatile from water, but some oxidants are GHS Category 4 toxicants and LD_{50} values were not available for several compounds (Table 3). In general, environmental contamination impacts are considered to be minor, but transportation and industrial use impacts are of interest for this group.

Glycol ethers and copper compounds are also used as breakers, though specific formulations and associated CAS numbers could not be located [16]. Exposure to low levels of glycol ethers may cause respiratory irritation, headache, and nausea, while exposure to high levels may have more severe health impacts, including liver and kidney damage, pulmonary edema, and loss of consciousness [84]. Inhalation of the dusts of copper salts may cause nasal and respiratory irritation, while contact with skin can cause itching, redness, or swelling [40]. Copper is a potent inhibitor of algae and is a toxic heavy metal that could represent a long-term environmental risk. Lack of information on specific formulations being used prevents a full assessment of breakers as a group of HF chemicals. The use of copper compounds, glycol ethers, and strong oxidants as breakers suggests that further investigation of the use of these chemicals in unconventional oil and gas recovery is warranted.

pH Adjusters

Acids and bases are added to fracturing fluids to adjust pH and improve effectiveness of certain chemical additives, particularly crosslinked polymer molecules. Ideal pH for a gelled treatment depends on the polymer being crosslinked and the crosslinking additive being used. Some crosslinkers perform better at high pH, such as borate crosslinking of guar and HPG, because a high pH is necessary for the crosslinking ion to dissociate from the crosslinking compound (due to high pK_a) [3]. When crosslinking HEC and HPC with zirconium, a pH between 10 and 12 is ideal; while CMHEC crosslinked with zirconium requires a pH between 4 and 6 [3].

In addition to optimizing crosslinker performance, pH control can be used as a method to manage bacterial growth in the fracturing fluid [85]. Acids are also used to clean the wellbore and dissolve formation minerals to open up flow paths for better fracturing near the wellbore [81] and [86]. Acids added to clean out the wellbore are typically not added directly to the fracturing fluid, but are instead used in conjunction with corrosion inhibitors and iron controllers between well perforation and HF [81]. Acid solutions used to dissolve carbonates or silicates for the purpose of increasing formation porosity (acidizing or acid-matrix treatments) are not considered in this study.

Typical pH adjusters include acetic acid, potassium hydroxide, sodium hydroxide, sodium carbonate, and potassium carbonate. Fumaric acid, previously discussed in section 3.1, is also employed as a pH adjuster. These pH adjusters are highly soluble in water and consist of biodegradable organic acids or ionic bases. Typical concentrations for pH adjusters range from 100 to 300 mg L^{-1}[5] and [63]. Some concentrated pH adjusters are hazardous, but many are of low toxicity (Table 3) and when diluted, they pose a minimal risk to the environment or human health. Sodium hydroxide and potassium hydroxide are the most toxic, considered GHS Category 3 toxic chemicals with oral rat LD_{50} values of 140 and 273 mg kg^{-1}, respectively. All other pH adjusters fall under Categories 4 and 5. Fumaric and acetic acids are expected to contribute to the COD of flowback water and affect oxidative treatments (Table 7). High or low pH solutions may need to be neutralized prior to treatment, as release of untreated high or low pH solutions can have negative environmental impacts or disrupt treatment processes, especially biological treatment trains. All strongly acidic or basic solutions are potentially hazardous to the environment and to human health and are a significant disposal problem, but procedures for the safe management of these solutions are available and are standard industry practice.

Biocides

Biocides are used in HF to control bacteria that degrade fracturing chemicals and contribute to corrosion of well tubing, casings, and equipment [16] and [81]. Sulfate reducing and acid forming bacteria are the primary organisms targeted with the application of biocides. Biocides frequently used for HF (Table 8) include quaternary ammonium compounds (QACs), glutaraldehyde, tetrakis hydroxymethyl phosphonium sulfate (THPS), tributyl tetradecyl phosphonium chloride (TTPC), and brominated compounds including 2,2-dibromo-3-nitrilopropionamide (DBNPA) [16] and [32]. Ammonium chloride, discussed in Section 3.3, is also used [28]. Biocides are typically used in concentrations ranging from 10 to 800 mg L^{-1}[5], [10],[63] and [81].

Quaternary ammonium compounds (QAC) are composed of straight chained alkyls and cyclic hydrocarbons surrounding a central nitrogen atom (Fig. 1f) [87]. The type of organic group and the chain length of the tail can vary, resulting in a range of properties for different QACs, including widely varying $\log K_{OW}$ and toxicity values [88]. Each CAS number for QACs corresponds to a specific blend of QAC molecules with specific attached alkyl groups [55]. Didecyl dimethyl ammonium chloride (DDAC) and alkyl dimethyl benzyl ammonium chloride (ADBAC) are commonly used QACs (Fig. 1g and h, respectively), and the physical and chemical constants for these compounds are included in Table 8. Toxicity information was found for DDAC, but not ADBAC (Table 3). The reported mammalian LD_{50} values for DDAC vary, but this compound is likely a GHS Category 3 toxin.

The physical properties of the QACs indicate that they will not volatilize from solution, and are likely to sorb to soils and sediments and will not migrate directly with groundwater (Table 8). QACs are common ingredients of household products and may be found in environmental samples impacted by septic and sewage systems [88] and [89]. The degree of biodegradability of QACs is variable. Data suggests that biodegradation decreases as the hydrophobic alkyl chain length increases and QACs that contain a benzyl group have lower biodegradation rates [88]. However, QACs have also been measured in sediments and waters near wastewater discharge sites, suggesting that QACs are environmentally persistent [89] and [90]. Measurements of QACs in surface waters and sediments has confirmed preferential sorption onto sediments, although aqueous concentrations were still detectable [89].

Glutaraldehyde and the phosphonium based biocides are sometimes considered "green" alternatives to other biocides, largely because they are less persistent in the environment. Glutaraldehyde is readily biodegradable under aerobic and anaerobic conditions [91]. The estimated aerobic and anaerobic half lives in river sediments are 10.6 and 7.7 h, respectively, based on measurements of glutaraldehyde losses[92], though conditions such as concentration and acclimation of bacteria to glutaraldehyde may affect degradation rates [93]. Glutaraldehyde has a moderate K_{OC} value, suggesting it will be mobile in soil and groundwater. Glutaraldehyde is volatile based on vapor pressure and meets criteria for GHS Category 1 toxic chemicals, according to some sources, based on rat inhalation studies (Table 2), suggesting the atmospheric contamination pathway should be considered. Glutaraldehyde is also of concern for oral toxicity (Table 3). Although glutaraldehyde is generally well-characterized and is not an environmentally persistent compound, the acute toxicity of this compound suggests that its use in HF should be further studied.

Tetrakis hydroxymethyl phosphonium sulfate (THPS) has low log K_{OW}, K_{OC}, and K_H values, suggesting that it will not sorb to soil or volatilize from solution, but will migrate with groundwater (Table 8). Under abiotic conditions, THPS has been shown to be stable at pH 5 and 7, with half-lives greater than 30 days; at pH greater than 8, THPS degrades within seven days. Under environmental conditions, THPS breaks down through hydrolysis, oxidation, and photodegradation [94] and initially degrades to trihydroxymethyl phosphine (THP), releasing two formaldehyde molecules and one sulfuric acid molecule in the process[56]. THPS is biodegradable, with carbon dioxide, water, and inorganic phosphate as the reported products[95]. Despite having rat oral LD_{50} values that place it as a GHS Category 3 toxin, THPS is considered a green alternative biocide due to its rapid biodegradation, low concentrations of harmful chemical intermediaries formed during degradation [95], low treatment dose levels, lack of halogens, and the lack of bioaccumulation potential [94] and [96].

Tributyl tetradecyl phosphonium chloride (TTPC) is a quaternary phosphonium biocide with a tetradecyl chain which causes the molecule to have surface-active properties [97]. Based on the oral

LD_{50} for rats, TTPC at a solution concentration of 48–52% is a GHS Category 4 chemical (Table 3). Chemical and physical data as well as information regarding the environmental fate and degradation of TTPC is not readily available for this compound, indicating TTPC should be further investigated.

2,2-Dibromo-3-nitrilopropionamide (DBNPA) is of moderate concern based on oral toxicity, as it is a GHS Category 3 chemical, but meets criteria for GHS Category 1 toxic chemicals based on rat inhalation LD_{50}values. However, DBNPA is unlikely to significantly volatilize from water, based on the Henry's constant, but it may volatilize from dry soils based on the vapor pressure. DBNPA has low log K_{OW}, and K_{OC}, suggesting that it will not sorb to soil and may migrate with groundwater (Table 8). Through hydrolysis, DBNPA degrades to dibromoacetonitrile, followed by dibromoacetomide, dibromoacetic acid, glyoxylic acid, and oxalic acid. The most stable of these products is dibromoacetic acid, with a half-life of approximately 300 days at 25 °C and pH 7.4 [98]. Degradation through contact with sunlight or reaction with nucleophiles results in formation of cyanoacetamide and ultimately cyanoacetic acid, malonic acid, and oxalic acid [98]. Biocide mixtures containing DBNPA typically contain a small percentage of 2-bromo-3-nitrilopropionamide, a known degradation product of DBNPA [98], [99] and [100]. Only limited chemical and physical information for 2-bromo-3-nitrilopropionamide is available.

1-Bromo-3-chloro-5,5-dimethylhydantoin has a broad range of reported mammalian oral LD_{50} values, generally in GHS Category 3 (Table 3), suggesting it is less toxic that DBNPA. Moderate volatilization from water is expected and it may be highly mobile in soils and groundwater based on a K_{OC} value of 23. When exposed to water, 1-bromo-3-chloro-5,5-dimethylhydantoin dissociates into dimethylhydantoin (DMH) and hypohalite ions [101]. The production of hypohalite ions has led to the use of 1-bromo-3-chloro-5,5-dimethylhydantoin for disinfection in water purification plants and cooling water systems, as well as residential applications such as swimming pools and hot tubs [101] and [102]. However, DMH may be of concern for environmental contamination due to its hydrolytic and photolytic stability [101]. The presence of biocides in HF wastewater may limit the treatment options by interfering with biological treatment methods or limit co-disposal of HF flowback with municipal wastewater. Studies have shown that QACs are not completely removed in wastewater treatment plants, and that at influent concentrations between 0.5 and 1 mg L^{-1} for benzalkonium chloride and 5 mg L^{-1} for dodecyl dialkyl ammonium chloride, these QACs can inhibit basic carbon respiration processes. Above 10 mg L^{-1} of either QAC, sludge nitrification can be inhibited [89]. In a study of the chemicals used in leather tanning, glutaraldehyde concentrations of 0.125–2.5 mg L^{-1} inhibited the activated sludge treatment process, with increasing inhibition occurring as the concentration increased, with nearly complete inhibition occurring at a concentration of 2.5 mg L^{-1}[103]. Given the wide use of biocides in HF, their diverse chemistry, their toxicity, and the lack of complete information on toxicology and use, biocides warrant additional investigation. More research is needed to determine the environmental fate and toxicity of QACs and other biocides, including biocides marketed as green alternatives.

Corrosion Inhibitors

Corrosion inhibitors are added to acid treatments and to fracturing fluids to form a protective layer on metal well components, preventing corrosion by acids, salts, and corrosive gasses [104], [105] and [106]. Selection of corrosion inhibitors and corresponding doses is dependent on well temperature, formation minerals, flow regime, and contact time. Typical doses of corrosion inhibitors range in concentrations from 10 to 7000 mg L^{-1}[5], [10], [63] and [81]. A variety of proprietary blends of chemicals are used; however, corrosion inhibitor mixtures often include acetaldehyde, acetone, ethyl methyl derivatives, formic acid, *n,n*-dimethyl formamide, propargyl alcohol, pyridinium, and thiourea (Table 7) [16], [32] and [105]. Isopropanol and methanol are also used as corrosion inhibitors and are discussed in Section 3.1. In general, corrosion inhibitors are highly soluble and biodegradable (Table

7). The low log K_{OW} and K_{OC}values indicate that these chemicals are not likely to sorb to soils, but that there is potential for these chemicals to migrate into surface and groundwater if released into the environment. Chemical data for ethyl methyl derivatives was not found because the specific compounds used and CAS numbers were not available [16]. Data for pyridinium (Fig. 1i) was also not available, and further investigation is recommended.

Corrosion inhibitors include chemicals that are toxic and carcinogenic. Acetaldehyde and thiourea are "reasonably anticipated to be a human carcinogen[s]" according to the U.S. Department of Health and Human Services National Toxicology Program [107]. Median lethal doses for corrosion inhibitors range from 20 to 9800 mg kg^{-1} via the oral route in rats (Table 3). Of the compounds for which toxicity data was found, propargyl alcohol and thiourea are GHS Category 2 chemicals, making them among the most toxic chemicals used in HF. Propargyl alcohol is considered readily biodegradable and although it is highly mobile in soil and groundwater, it is not expected to be environmentally persistent. Thiourea is only considered inherently biodegradable and is expected to be highly mobile in soil and groundwater, increasing its potential for environmental impacts. Due to environmental regulations and concerns about toxicity, development is underway for new oxygen-based corrosion inhibitors and other substitutes that are biodegradable, less toxic, and non-bioaccumulating [108]. Corrosion inhibitors are widely used in industry, and as a group they are not well characterized (Table 7), and they deserve further investigation based on their toxicity.

Scale Inhibitors

Scale inhibitors, including phosphonic acid salts, sodium polycarboxylate, and copolymers of acrylamide and sodium acrylate, are added to fracturing fluids to protect piping and prevent formation plugging [32],[109] and [110]. Scaling reduces well production by blocking formation pores, reducing permeability, and by blocking flow in piping and tubing. Scale inhibitors are typically used in low concentrations, ranging from 75 to 400 mg L^{-1} in fracturing fluid [5], [10], [63] and [81]. Most scale inhibitors work by blocking nucleation sites on growing mineral crystals, preventing further growth [109]. Polycarboxylates and acrylate polymers are the most commonly used additives, as phosphonate compounds are often incompatible with other fracturing fluid additives [81]. To our knowledge, no public chemical data is available on the identified scale inhibitors or could not be found due to the lack of CAS numbers or chemical forumlas reported in the FracFocus database, thus, further investigation of these chemicals is recommended.

Iron Control

Chemicals may be added to fracturing fluids to control iron precipitates that block flow paths within the formation, reducing reservoir rock permeability, well productivity, and fluid recovery [111] and [112]. Ferric iron (Fe^{3+}) can also inadvertently act as a crosslinker in fracturing fluids containing gelling agents, changing the viscosity of the fluid [111]. Iron precipitation is prevented through the use of iron controlling agents such as thioglycolic acid, citric acid, acetic acid, and sodium erythorbate [9]. Iron controlling agents are typically used in a concentration of 50–200 mg L^{-1}[5], [10] and [63]. These chemicals work by acting as chelating agents, forming complexes with ferrous iron (Fe^{2+}) to prevent the iron from precipitating[81] and [113]. The use of iron control agents depends on the shale characteristics; Marcellus shale usually does not contain enough iron to warrant iron controlling agents, while the Fayetteville shale does contain high quantities of iron [9].

Acetic acid, citric acid, and sodium erythorbate are common food additives, and as a result, they are of low toxicity to humans. Acetic acid and citric acid are readily biodegradable; however, no data concerning the biodegradability of sodium erythorbate was found. Thioglycolic acid is commonly used in hair care products and leather processing, and while biodegradable, poses a greater toxicity risk due to an oral LD_{50} value of 114 mg kg^{-1} in rats, classifying it as a GHS Category 3 toxic chemical.

All four iron-controlling agents investigated are highly soluble in water and are expected to contribute to the COD of flowback water (Table 7). With the exception of acetic acid, the chemicals are not expected to volatilize from fracturing fluid based on Henry's constants (Table 7). The low K_{OC} values of citric acid and thioglycolic acid indicate that these chemicals will not sorb extensively to soils, but will be mobile in surface and groundwater (Table 7). The K_{OC} of acetic acid is higher and is pH dependent, with a range from 6.5 to 228, but is still relatively low (Table 7). Of the iron control agents, thioglycolic acid appears to be the most problematic as a potential environmental contaminant, although all are generally readily degraded and not persistent.

Clay Stabilizers

Clay stabilizers are used to prevent the swelling of clays found in gas shale layers, particularly smectite. Some gas-containing shales can contain up to 50% clays by volume [15]. Clay swelling and migration can cause borehole instability and can reduce reservoir rock permeability by up to 90%, reducing well productivity and contributing to complications such as sticking of the drill-pipe in the borehole[114] and [115]. The degree of clay swelling is dependent on the salinity of the fracturing fluid and the species of cations present, with higher salinities generally reducing swelling [114]. Among cations, divalent cations generally result in less swelling than monovalent cations, and potassium and ammonium salts contain the preferred monovalent cations [114]. Clay migration is also an issue, as clays can become suspended and dispersed by produced water or fracturing fluid, clogging pores and reducing productivity[116]. Clay stabilizers work via ion exchange, replacing the cations (such as sodium) in the clay with other cations (such as divalent cations) that have a lesser tendency to become hydrated and swell the clay. Fracturing fluid that is in contact with clay must maintain a minimum concentration of clay stabilizers to prevent a reversal in cation exchange and clay swelling [114]. Clay stabilizers typically comprise 0.05–0.2% of the total fracturing fluid by volume and concentrations range from 500 to 2000 mg L^{-1}[5], [10], [63] and [81]. Clay stabilizers include choline chloride, tetramethyl ammonium chloride, potassium chloride, and sodium chloride [32]. Potassium chloride and sodium chloride are widely used in industry and are well known, and are therefore not considered high priority compounds for concern, however saline wastewaters can present significant management challenges.

Tetramethyl ammonium chloride, a QAC used as a clay stabilizer rather than as a biocide, is a GHS Category 2 oral toxin (Table 3) and one of the most toxic chemicals identified in this study. Tetramethyl ammonium chloride poses an environmental risk as it has been deemed non-biodegradable in aerobic studies and is expected to be highly mobile in soils based on K_{OC} values (Table 9). Although, tetramethyl ammonium chloride is not expected to volatilize from water, if dispersed into the air, any volatile fraction will rapidly degrade via hydroxyl radicals [40]. Although tetramethyl ammonium chloride is expected to contribute to fracturing fluid COD, it is not necessarily biodegradable. Tetramethyl ammonium chloride is also used as a catalyst for organic synthesis and polymerase chain reaction [117], though reports of other uses are limited. Due to the fact that tetramethyl ammonium chloride is one of the few GHS Class 2 toxins identified as being used in HF and it is in the same chemical group as QAC biocides, this compound should be further investigated and its use in HF applications better understood.

Choline is readily biodegradable and generally poses minor or no health risk to humans, as it is falls in GHS Category 5. Choline chloride is used as a food additive and choline is considered an essential nutrient in humans [118]. Choline chloride is expected to be non-volatile from water and a contributor to fracturing fluid COD (Table 9). Choline chloride is compatible with most fracturing fluids, breakers, and crosslinkers, making it an attractive clay stabilizer [119]. Current trends towards the use of choline chloride, as opposed to ammonium salts and the historically used sodium chloride and potassium chloride, are meant to reduce environmental impacts through increased biodegradability, lower toxicity, and lower concentrations in fracturing fluid, as choline chloride is more effective at reducing clay swelling at lower concentrations [81],[86] and [114].

Surfactants

Surfactants are used in HF fluid to control for optimal viscosity of fracturing fluids, reduce surface tension between the shale formations and the HF fluid, and assist fluid recovery after fracturing [3] and [81]. In some instances, surfactants may also act as biocides or clay stabilizers [3]. Surfactants are important in emulsion-based fracturing fluids that contain both hydrocarbon and aqueous phases, ensuring proper viscosity and transport properties [3]. Surfactants also reduce the wetting of shale interfaces, allowing for increased gas flow through channels in the formation.

Surfactants, more specifically viscoelastic surfactants, can also be used in place of crosslinkers and gelling agents and foaming agents in high temperature and pressure formations to maintain fluid viscosity while reducing formation damage and increasing fluid recovery [120]. Viscoelastic surfactants form rod-shaped micelles that associate with each other in fracturing fluids containing proper concentrations of salts, resulting in high viscosity and elasticity [3] and [86]. However, when viscoelastic surfactants come into contact with formation water or hydrocarbons their structure is disrupted, lowering viscosity without the need for breakers while leaving minimal residue in the formation [3].

Surfactants can vary greatly, consisting of amphoteric, anionic, or non-ionic compounds [81]. Typical surfactants include sodium lauryl sulfate and dimethyl dihydrogenated tallow ammonium chloride. Product stabilizers and winterizing agents, mistakenly identified as surfactants in FracFocus, including ethanol, 2-butoxyethanol, and isopropanol are discussed above [32]. Depending on their intended purpose, surfactants are typically used in concentrations ranging from 500 to 1800 mg L^{-1} in fracturing fluid [5], [10],[63] and [81]. No chemical data concerning dimethyl dihydrogenated tallow ammonium chloride was available in the databases surveyed due to the lack of a CAS number and exact chemical composition, and thus, further investigation on this compound is recommended. Generally, the surfactants investigated are highly soluble in water, readily biodegradable, and are expected to contribute to flowback COD. Sodium lauryl sulfate has a moderately high K_{OC} value, and is expected to have moderate to low mobility in soil and groundwater [40] and [41]. Sodium lauryl sulfate is found in household products and is not anticipated to be a health risk due to its LD_{50} value.

CONCLUSIONS

Eighty-one commonly used HF fluid additives were identified from the FracFocus database and other sources and examined. The chemicals used in HF treatments function as gelling and foaming agents, friction reducers, crosslinkers, breakers, pH adjusters, biocides, corrosion inhibitors, scale inhibitors, iron control chemicals, clay stabilizers, and surfactants. Not all compounds are used in all treatments; for example, gels are not typically used with friction reducers. Within each functional category there are choices of chemicals that can be used; for example, there are several classes of biocides that could potentially be used interchangeably.

Fifty-five of the HF chemicals identified were organic, with 27 of these considered readily or inherently biodegradable according to OECD guidelines [60]. Four chemicals had high CODs (>3 gO_2 g^{-1}) (monoethyl amine, pyridinium, didecyl dimethyl ammonium chloride, tributyl tetradecyl phosphonium chloride), while another 13 had moderate CODs (>2 gO_2 g^{-1}), and many of these compounds are used at significant concentrations in HF treatments. Using data from Table 4, Table 5, Table 6, Table 7, Table 8 and Table 9 and the reported concentrations at which HF chemicals are used in HF treatments [5],[10] and [63], we estimate that the COD of HF fluids will be in the range of 5000 mg L^{-1} or more, with potential values ranging four times higher if high concentrations of corrosion inhibitors are used. These calculated values correspond reasonably well with reported values for flowback water [1],[18] and [19], and it can be concluded that flowback will be a high COD wastewater, but there are very few studies characterizing flowback and further investigation of flowback water quality is needed. High COD industrial wastes present significant treatment challenges. Further study of individual chemicals, chemical mixtures, and actual flowback are needed to determine what treatment technologies, including aerobic and anaerobic biological treatment, are appropriate for HF wastewaters. The use of some membrane treatment technology may be problematic given the high COD and high expected biodegradability of HF waste waters.

Overall, volatility of the chemicals examined was not a major concern, as only 12 chemicals were considered volatile or semi-volatile from water based on Henry's constants (Table 4, Table 5, Table 6, Table 7, Table 8 and Table 9). However, given that some HF chemicals meet criteria for GHS Category 1 and 2 inhalation toxins (Table 2), the potential significance of the volatile exposure pathway cannot be ignored and should be further investigated.

Overall, most of the chemicals that had available toxicity data are of moderate or little concern based on mammalian acute oral toxicity. None of the HF chemicals were classified as GHS Category 1 oral toxins; however, three chemicals were classified as GHS Category 2 oral toxins (propargyl alcohol, thiourea, tetramethyl ammonium chloride) and ten were classified as GHS Category 3 oral toxins (Table 3). Importantly, there remains a significant gap in toxicity information, as no mammalian toxicity data was found for approximately one-third of the 81 chemicals examined. Poorly characterized compounds include biocides, corrosions inhibitors (e.g. pyridinium), scale inhibitors, and iron control agents which are widely used in industry. Biocides as a group are of concern, as they contain the highest number of toxic compounds. Other identified chemicals of concern, based on inhalation toxicity, are glutaraldehyde and thioglycolic acid (Table 2). Furthermore, at least five chemicals are confirmed or suspected carcinogens (ethanol, naphthalene, diethanolamine, acetaldehyde, thiourea).

While increased chemical disclosure by companies that perform HF is an important step towards understanding the potential risks associated with contamination events, organization of that information is still required for a more comprehensive understanding of potential environmental and health hazards of chemicals used in unconventional oil and gas extraction processes. HF contractors and companies conducting these operations should continue to disclose information to databases such as FracFocus, but more information about the compositions of proprietary additives and chemical derivatives used in fracturing processes, concentrations at which chemical are used, and mass amounts of chemicals used would be useful for environmental risk assessment.

All of the chemicals used in unconventional production are widely used in industry and data gaps concerning toxicity, biodegradability, physical constants, and concentrations of use should be addressed so that accurate and informed environmental and health assessments can be made. It is recommended that log K_{ow}, Henry's law constant, and other fundamental physical and chemical properties be determined and reported for each compound [59]. In order to understand the environmental fate of HF chemicals during accidental or deliberate releases, tests on biodegradability of pure compounds and mixtures, under both aerobic and anaerobic conditions, should be conducted [60]. At a minimum, acute mammalian oral toxicity, using rats or mice and protocols consistent with international standards [61] and [121], should be measured and reported for each compound. Environmental toxicity information, using tests on standard fish, crustacean, and plant or algae species [121], is recommended to support complete environmental assessments of industrial chemicals.

Finally, there is a crucial need for more field studies, including characterization of HF fluids before injection and characterization of flowback after injection. Studies need to be conducted to examine the fate of HF chemicals in complex environmental systems with the objective of more completely understanding environmental hazards associated with HF chemicals and chemical mixtures. These studies need to include an evaluation of complex mixtures and examine how the interactions between chemicals found in HF fluids influences the environmental fate of individual components.

ACKNOWLEDGMENTS

The work was completed by the Ecological Engineering Research Program with funding from the University of the Pacific, School of Engineering and Computer Science. Part of this work was conducted at Lawrence Berkeley National Laboratory under its U.S. Department of Energy contract DE-AC02-05CH11231.

REFERENCES

1. New York State Department of Environmental Conservation (NYS DEC), Revised Draft Supplemental Generic Environmental Impact Statement on the Oil, Gas and Solution Mining Regulatory Program, Well Permit Issuance for Horizontal Drilling and High-Volume Hydraulic Fracturing in the Marcellus Shale and Other Low-Permeability Gas Reservoirs, New York State Department of Environmental Conservation (NYS DEC), 2011, http://www.dec.ny.gov/energy/75370.html(accessed July 1, 2013).
2. L. Britt, Fracture stimulation fundamentals,J. Nat. Gas Sci. Eng. 8 (2012) 34–51.
3. M.J. Economides, K.G. Nolte, Reservoir Stimulation, third ed., John Wiley & Sons, Chichester, England, 2000.
4. G. Zimmermann, G. Blocher, A. Reinicke, W. Brandt, Rock specific hydraulic fracturing and matrix acidizing to enhance a geothermal system—concepts and field results, Tectonophysics 503 (2011) 146–154.
5. Ground Water Protection Council, ALL Consulting, Modern Shale Gas Development in the United States: A Primer, U.S. Department of Energy, Office of Fossil Energy, Washington, DC, 2009.
6. International Energy Agency (IEA), Golden Rules for a Golden Age of Gas: World Energy Outlook Special Report

on Unconventional Gas, WEO-2012, International Energy Agency (IEA), Paris, France, 2012.

7. U.S. Energy Informations Administration (EIA), Annual Energy Outlook 2012 with projections to 2035, in: DOE/EIA-0383(2012), U.S. Energy Informations Administration (EIA), Washington, DC, 2012.
8. International Energy Agency (IEA), World Energy Outlook 2012: Executive Summary, International Energy Agency (IEA), Paris, France, 2012.
9. International Energy Agency (IEA), Water Management Associated with Hydraulic Fracturing, AP Guidance Document HF2, International Energy Agency (IEA), Washington, DC, 2010.
10. G.E. King, Hydraulic fracturing 101: What every representative, environmentalist, regulator, reporter, investor, university researcher, neighbor and engineer should know about estimating frac risk and improving frac performance in unconventional gas and oil wells, in: SPE Hydraulic Fracturing Technology Conference, February 6–8, 2012, The Woodlands, TX, Society of Petroleum Engineers, 2012.
11. A.L. Maule, C.M. Makey, E.B. Benson, I.J. Burrows, M.K. Scammell, Disclosure of hydraulic fracturing fluid chemical additives: analysis of regulations, New Solution: J. Environ. Occup. Heal. Pol. 23 (2013) 167–187.
12. Air Products, Enhanced Unconventional Oil and Gas Production with Nitrogen and Carbon Dioxide Fracturing, 351-13-006-US, 2013, Air Products and Chemicals; Allentown, PA.http://www.airproducts.com/industries/ Energy/oilgas-production/oilfield-services/product-list/~/media/086A940DF83843A89D921BD12F0454BD.pdf(accessed April 1, 2014).
13. O. Hoch, The dry coal anomaly—the Horseshoe Canyon Formation of Alberta, Canada, in: Society of Petroleum Engineers Annual Technical Conference and Exhibition,

October 9–12, 2005, Dallas, TX, Society of Petroleum Engineers, 2005.

14. D.M. Kargbo, R.G. Wilhelm, D.J. Campbell, Natural gas plays in the Marcellus Shale: challenges and potential opportunities, Environ. Sci. Technol. 44 (2010) 5679–5684.
15. P. Kaufman, G. Penny, J. Paktinat, Critical evaluation of additives used in shale slickwater frac, in: 2008 SPE Shale Gas Production Conference, November 16–18, 2008, Irving, TX, Society of Petroleum Engineers, 2008.
16. U.S. Environmental Protection Agency (U.S. EPA), Chapter 4: hydraulic fracturing fluids, in: Evaluation of Impacts to Underground Sources of Drinking Water by Hydraulic Fracturing of Coalbed Methane Reservoirs, EPA 816-R- 04-003, U.S. Environmental Protection Agency (U.S. EPA), Washington, DC, 2004.
17. K.E. Cawiezel, Delayed crosslinking system for fracturing fluids, Pumptech N.V. & Compagnie Des Services Dowell Schlumberger, in: EPO Patent 0347975 A2, 1989.
18. E. Barbot, N.S. Vidic, K.B. Gregory, R.D. Vidic, Spatial and temporal correlation of water quality parameters of produced waters from devonian-age shale following hydraulic fracturing, Environ. Sci. Technol. 47 (2013) 2562–2569.
19. L.O. Haluszczak, A.W. Rose, L.R. Kump, Geochemical evaluation of flowback brine from Marcellus gas wells in Pennsylvania, USA, Appl. Geochem. 28 (2013) 55–61.
20. K.B. Gregory, R.D. Vidic, D.A. Dzombak, Water management challenges associated with the production of shale gas by hydraulic fracturing, Elements 7 (2011) 181–186.
21. B.R. Hansen, S.R.H. Davies, Review of potentialtechnologies for the removal of dissolved components from produced water, Chem. Eng. Res. Des. 72 (1994) 176–188.
22. J.A.Veil, C.E. Clark, Produced-water-volume estimates and management practices, Soc. Pet. Eng. Prod. Oper. 26 (2011) 234–239.

23. H. Hatzenbuhler, T.J. Centner, Regulation of water pollution from hydraulic fracturing in horizontally-drilled wells in the Marcellus Shale Region, USA, Water 4 (2012) 983–994.
24. B.G. Rahm, S.J. Riha, Toward strategic management of shale gas development: regional, collective impacts on water resources, Environ. Sci. Policy 17 (2012) 12–23.
25. D.J. Rozell, S.J. Reaven,Water pollution risk associated with natural gas extraction from the Marcellus Shale, Risk Anal. 32 (2012) 1382–1393.
26. U.S. Environmental Protection Agency (U.S. EPA), Plan to study the potential impacts of hydraulic fracturing on drinking water resources, in: EPA/600/R- 11/122, U.S. Environmental Protection Agency (U.S. EPA), Washington, D.C., 2011.
27. R.D. Vidic, S.L. Brantley, J.M. Vandenbossche, D. Yoxtheimer, J.D. Abad, Impact of shale gas development on regional water quality, Science 340 (2013), http://dx.doi.org/10.1126/science.1235009.
28. A. Aminto, M.S. Olson, Four-compartment partition model of hazardous components in hydraulic fracturing fluid additives, J. Nat. Gas Sci. Eng. 7 (2012) 16–21.
29. S.A. Gross, H.J. Avens, A.M. Banducci, J. Sahmel, J.M. Panko, B.E. Tvermoes, Analysis of BTEX groundwater concentrations from surface spills associated with hydraulic fracturing operations, J. Air Waste Manage. 63 (2013) 424–432.
30. B.G. Rahm, J.T. Bates, L.R. Bertoia, A.E. Galford, D.A. Yoxtheimer, S.J. Riha, Wastewater management and Marcellus Shale gas development: trends, drivers, and planning implications, J. Environ. Manage. 120 (2013) 105–113.
31. D. Rahm, Regulating hydraulic fracturing in shale gas plays: the case of Texas, Energy Policy 39 (2011) 2974–2981.
32. FracFocus, FracFocus Chemical Disclosure Registry, Ground Water Protection Council and Interstate Oil and Gas Compact Commission, 2013, http://fracfocus.org(accessed March 8, 2013).

33. FracFocus, What Chemicals are Used? Ground Water Protection Council and Interstate Oil and Gas Compact Commission, 2013, http://fracfocus.org/chemical-use/what-chemicals-are-used(accessed March 8, 2013).
34. SkyTruth, Fracking Chemical Database, 2013, http://frack.skytruth.org/fracking-chemical-database(accessed June 2013).
35. S. Gartiser, E. Urich, Elimination of cooling water biocides in batch tests at different inoculum concentrations, in: Society of Environmental Toxicology and Chemistry Europe 13th Annual Meeting, April 28–May 1, 2003, Hamburg, Germany, 2003.
36. T. Nishihara, T. Okamoto, N. Nishiyama, Biodegradation of didecyldimethylammonium chloride by pseudomonas fluorescens TN4 isolated from activated sludge, J. Appl. Microbiol. 88 (2000) 641–647.
37. J.D. Arthur, B. Bohm, M. Layne, Hydraulic Fracturing Considerations for Natural Gas Wells of the Marcellus Shale, The Ground Water Protection Council, Cincinnati, OH, 2008.
38. J. Miller, Biodegradable surfactants aid the development of environmentally acceptable drilling fluid additives, in: International Symposium on Oilfield Chemistry, February 28–March 2, 2007, Houston, TX, Society of Petroleum Engineers, 2007.
39. Halliburton Energy Services, GBW-30 Breaker MSDS, Halliburton Energy Services, Duncan, OK, 2009.
40. National Library of Medicine, Toxicology Data Network (TOXNET) Hazardous Substance Data Bank (HSDB), National Library of Medicine, 2013, http://toxnet.nlm.nih.gov/cgi-bin/sis/htmlgen?HSDB(accessed June 2013).
41. European Chemicals Agency (ECHA), International Uniform Chemical Information Database (IUCLID), CD-ROM Year 2000 Edition, European Chemicals Agency (ECHA), 2000.

42. Syracuse Research Corporation (SRC), Physical Properties Database (PhysProp), 2011, http://www.srcinc.com/what-we-do/database forms.aspx?id=386(accessed June 2013).

43. Knovel, Knovel Critical Tables, second ed., Knovel, 2008, http://app. knovel.com/hotlink/toc/id:kpKCTE000X/knovel-critical-tables(accessed March 2013).

44. U.S. Environmental Protection Agency (U.S. EPA), ACToR (Aggregated Computational Toxicology Resource) Database, U.S. Environmental Protection Agency (U.S. EPA), 2013, http://actor.epa.gov/actor/faces/ACToRHome.jsp (accessed March 2013).

45. National Center for Biotechnology Information, PubChem Database, National Library of Medicine, National Center for Biotechnology Information, 2013, http://pubchem.ncbi.nlm.nih.gov (accessed March 2013).

46. National Library of Medicine, ChemIDplus Advanced. http://chem. sis.nlm.nih.gov/chemidplus/(accessed February 2014).

47. W.M. Haynes, D.R. Lide, T.J. Bruno, CRC Handbook of Chemistry and Physics 2012–2013, 93rd ed., CRC Press, Boca Raton, FL, 2012.

48. R.J. Watts, Hazardous Wastes: Sources, Pathways, Receptors, John Wiley & Sons, Inc., New York, NY, 1998.

49. P.H. Howard, W. Meylan, Handbook of Physical Properties of Organic Chemicals, CRC Press, Inc., Boca Raton, FL, 1997.

50. R.J. Lewis, N.I. Sax, Sax's Dangerous Properties of Industrial Materials, ninth ed., Van Nostrand Reinhold, New York, NY, 1996.

51. R.J. Lewis, Hawley's Condensed Chemical Dictionary, 13th ed., Van Nostrand Reinhold, New York, NY, 1997.

52. BioLab Water Additives, 1-Bromo-3-Chloro-5,5-Dimethylhydantoin (Bromicide®), Product Information, BioLab Water Additives, Manchester, UK, 1999.

53. BWA Water Additives, Bellacide 350 MSDS, BWA Water Additives, Tucker, GA, 2008.

54. C.J.Bradaric,A. Downard,C.Kennedy,A.J.Robertson,Y. Zhou,Industrial preparation of phosphonium ionic liquids, Green Chem. 5 (2003) 143–152.
55. U.S. Environmental Protection Agency (U.S. EPA), Reregistration eligibility decision for alkyl dimethyl benzyl ammonium chloride (ADBAC), in: EPA739- R-06-009, U.S. Environmental Protection Agency (U.S. EPA), Washington, DC, 2006.
56. A.N. Shamim, Scoping Document: Product Chemistry/ Environmental Fate/Ecotoxicity of: Tetrakis (Hydroxymethyl) Phosphonium Sulfate, U.S. Environmental Protection Agency (U.S. EPA), Washington, DC, 2011.
57. Organization for Economic Cooperation and Development (OECD), Screening Information Data Set (SIDS) for High Volume Chemicals, Organization for Economic Cooperation and Development (OECD), 2007, http://www.chem.unep.ch/irptc/sids/OECDSIDS/sidspub.html (accessed June 2013).
58. European Chemicals Agency (ECHA), SVHC Support Document, Disodium Tetraborate, Anhydrous, European Chemicals Agency (ECHA), 2010.
59. Organization for Economic Cooperation and Development (OECD), OECD Guidelines for the Testing of Chemicals, Section 1: Physical-Chemical Properties, Organization for Economic Cooperation and Development (OECD), 2013.
60. Organization for Economic Cooperation and Development (OECD), OECD Guidelines for the Testing of Chemicals, Section 3: Degradation and Accumulation, Organization for Economic Cooperation and Development (OECD), 2013.
61. United Nations, Globally harmonized system of classification and labelling of chemicals (GHS), in: ST/SG/AC.10/30/Rev.5, United Nations, New York and Geneva, 2013.
62. J.Y. Wang, S.A. Holditch, D.A. McVay, Effect of gel damage on fracture fluid cleanup and long-term recovery in tight gas reservoirs, J. Nat. Gas Sci. Eng. 9 (2012) 108–118.

63. URS Corporation, Water-related issues associated with gas production in the Marcellus Shale: Additives use, flowback quality and quantities, regulations, on-site treatment, green technologies, alternate water sources, water well-testing, in: Contract PO No. 10666, URS Corporation, Fort Washington, Pennsylvania, 2011.
64. T.N.C. Dantas, V.C. Santanna, A.A.D. Neto, M. Moura, Hydraulic gel fracturing, J. Dispersion Sci. Technol. 26 (2005) 1–4.
65. D. Mudgil, S. Barak, B.S. Khatkar, Effect of enzymatic depolymerization on physicochemical and rheological properties of guar gum, Carbohyd. Polym. 90 (2012) 224–228.
66. B. Zhang, A. Huston, L. Whipple, H. Urbina, K. Barrett, M. Wall, R. Hutchins, A. Mirakyan, A superior, high-performance enzyme for breaking borate crosslinked fracturing fluids under extreme well conditions, Soc. Pet. Eng. Prod. Oper. 28 (2013) 210–216.
67. Y. Cheng, R.K. Prud'homme, Enzymatic degradation of guar and substituted guar galactomannans, Biomacromolecules 1 (2000) 782–788.
68. H.H. Gong, M.Z. Liu, J.C. Chen, F. Han, C.M. Gao, B. Zhang, Synthesis and characterization of carboxymethyl guar gum and rheological properties of its solutions, Carbohyd. Polym. 88 (2012) 1015–1022.
69. P. Adhikary, S. Krishnamoorthi, R.P. Singh, Synthesis and characterization of grafted carboxymethyl guar gum, J. Appl. Polym. Sci. 120 (2011) 2621–2626.
70. R. Fujioka, Y. Tanaka, T. Yoshimura, Synthesis and properties of superabsorbent hydrogels based on guar gum and succinic anhydride, J. Appl. Polym. Sci. 114 (2009) 612–615.
71. Y. Lester, T. Yacob, I. Morrissey, K.G. Linden, Can we treat hydraulic fracturing flowback with a conventional biological process? The case of guar gum, Environ. Sci. Technol. Lett. (2013), http://dx.doi.org/10.1021/ez4000115.

72. B. Kwon, S. Lee, J. Cho, H. Ahn, D. Lee, H.S. Shin, Biodegradability, DBP formation, and membrane fouling potential of natural organic matter: Characterization and controllability, Environ. Sci. Technol. 39 (2005) 732–739.

73. L. Metcalf, H.P. Eddy, G. Tchobanoglous, F.L. Burton, D.H. Stensel, Wastewater Engineering, Treatment and Reuse, fourth ed., McGraw-Hill, New York, NY, 2003.

74. M.T. Bao, Q.G. Chen, Y.M. Li, G.C. Jiang, Biodegradation of partially hydrolyzed polyacrylamide by bacteria isolated from production water after polymer flooding in an oil field, J. Hazard. Mater. 184 (2010) 105–110.

75. Q.X. Wen, Z.Q. Chen, Y. Zhao, H.C. Zhang, Y.J. Feng, Biodegradation of polyacrylamide by bacteria isolated from activated sludge and oil-contaminated soil, J. Hazard. Mater. 175 (2010) 955–959.

76. L.L. Liu, Z.P. Wang, K.F. Lin, W.M. Cai, Microbial degradation of polyacrylamide by aerobic granules, Environ. Technol. 33 (2012) 1049–1054.

77. P.D. Moffitt, A. Moradi-Araghi, I. Ahmed, V.R. Janway, G.R. Young, Development and field testing of a new low toxicity polymer crosslinking system, in: Permian Basin Oil and Gas Recovery Conference, March 27–29, 1996, Midland, TX, Society of Petroleum Engineers, 1996.

78. R.H. Nielsen, G. Wilfing, Zirconium and zirconium compounds, in: Ullmann's Encyclopedia of Industrial Chemistry, Wiley-VCH Verlag GmbH & Co. KGaA, Weinheim, Germany, 2012, pp. 753–778.

79. H.D. Brannon, R.M. Tjon-Joe-Pin, Biotechnological breakthrough improves performance of moderate to high-temperature fracturing applications, in: Society of Petroleum Engineers Annual Technical Conference and Exhibition, September 25–28, 1994, New Orleans, LA, Society of Petroleum Engineers, 1994.

80. R. Barati, S.J. Johnson, S. McCool, D.W. Green, G.P. Willhite, J.T. Liang, Fracturing fluid cleanup by controlled release of

enzymes from polyelectrolyte complex nanoparticles, J. Appl. Polym. Sci. 121 (2011) 1292–1298.

81. R. McCurdy, High rate hydraulic fracturing additives in non-Marcellus unconventional shale, in: Proceedings ofthe Technical Workshops for the Hydraulic Fracturing Study: Chemical & Analytical Methods, February 24–25, 2011, Arlington, VA, February, U.S. Environmental Protection Agency, 2011.

82. J. Gulbis, M.T. King, G.W. Hawkins, H.D. Brannon, Encapsulated breaker for aqueous polymeric fluids, Soc. Pet. Eng. Prod. Eng. 7 (1992) 9–14.

83. P.V. Coveney, H.d. Silva, A. Gomtsyan, A. Whiting, E.S. Boek, Novel approaches to cross-linking high molecular weight polysaccharides: application to guarbased hydraulic fracturing fluids, Mol. Simul. 25 (2000) 265–299.

84. U.S. Environmental Protection Agency (U.S. EPA), Glycol Ethers (2-Methoxyethanol, 2-Ethoxyethanol, and 2-Butoxyethanol), U.S. Environmental Protection Agency (U.S. EPA), 2000, http://www.epa.gov/ ttnatw01/hlthef/glycolet.html(accessed August 10, 2013).

85. Halliburton, MO-67TM pH Control Agent Brochure, H02240 8/09, Halliburton, 2009, http://www.halliburton.com/public/pe/contents/ Chem Compliance/web/H02240.pdf(accessed March 8, 2013).

86. J. Fink, Petroleum Engineer's Guide to Oil Field Chemicals and Fluids, first ed., Gulf Professional Publishing, Waltham, MA, 2012.

87. D.L. Fredell,Biologicalproperties and applications of cationic surfactants, Surf. Sci. Ser. 53 (1994) 31.

88. M.T. Garcı a, I. Ribosa, T. Guindulain, J. Sánchez-Leal, J. Vives-Rego, Fate and effect of monoalkyl quaternary ammonium surfactants in the aquatic environment, Environ. Pollut. 111 (2001) 169–175.

89. N. Kreuzinger, M. Fuerhacker, S. Scharf, M. Uhl, O. Gans, B. Grillitsch, Methodological approach towards the

environmental significance of uncharacterized substances - Quaternary ammonium compounds as an example, Desalination 215 (2007) 209–222.

90. C.A.M. Bondi, Applying the precautionary principle to consumer household cleaning product development, J. Cleaner Prod. 19 (2011) 429–437.
91. L. Laopaiboon, N. Phukoetphim, K. Vichitphan, P. Laopaiboon, Biodegradation of an aldehyde biocide in rotating biological contactors, World J. Microb. Biot. 24 (2008) 1633–1641.
92. L.L. Sano, A.M. Krueger, P.F. Landrum, Chronic toxicity of glutaraldehyde: differential sensitivity of three freshwater organisms, Aquat. Toxicol. 71 (2005) 283–296.
93. H.W. Leung, Ecotoxicology of glutaraldehyde: review of environmental fate and effects studies, Ecotoxicol. Environ. Saf. 49 (2001) 26–39.
94. U.S. Environmental Protection Agency (U.S. EPA), Presidential Green Chemistry Challenge Award Recipients 1996–2012, 744F12001, U.S. Environmental Protection Agency (U.S. EPA), Washington, DC, 2012.
95. S. Groome, Tetrakis (hydroxymethyl) phosphonium sulfate THPS chemistry document, in: EPA-HQ-OPP-2011-0067, U.S. Environmental Protection Agency, Washington, DC, 2011, in press.
96. S.A. Drozdz, V.F. Hock, Green chemical treatments for heating and cooling systems, in: ERDC/CERL TR-06-29, U.S. Army Engineer Research and Development Center, Champaign, IL, 2006.
97. J.F. Kramer, F. O'Brien, S.F. Strba, A new high performance quaternary phosphonium biocide for microbiological control in oilfield water systems, in: Corrosion 2008 Conference & Expo, NACE International, New Orleans, LA, 2008.
98. J.H. Exner, G.A. Burk, D. Kyriacou, Rates and products of decomposition of 2,2-dibromo-3-nitrilopropionamide, J. Agric. Food Chem. 21 (1973) 838–842.

99. Dow Chemical Company, Material Safety Data Sheet: DBNPA 100 PTECH, Dow Chemical Company, 2003.
100. F.A. Blanchard, S.J. Gonsior, D.L. Hopkins, 2,2-Dibromo-3-nitrilopropionamide (DBNPA) chemical degradation in natural waters: Experimental evaluation and modeling of competitive pathways, Water Res. 21 (1987) 801–807.
101. U.S. Environmental Protection Agency (U.S. EPA), Reregistration Eligibility Decision for Halohydantoins (Case 3055), EPA 739-R-07-001, U.S. Environmental Protection Agency (U.S. EPA), Washington, DC, 2007.
102. C. Avendano, ˜ J.C. Menendez, Hydantoin and its derivatives, in: Kirk-Othmer Encyclopedia of Chemical Technology, John Wiley & Sons, Inc., New York, NY, 2000.
103. S. Danhong, H. Qiang, Z. Wenjun, W. Yulu, S. Bi, Evaluation of environmental impact of typical leather chemicals. Part II: Biodegradability of organic tanning agents by activated sludge, J. Soc. Leather Technol. Chem. 92 (2008) 59–64.
104. A.A. Al-Zahrani, Innovative method to mix corrosion inhibitor in emulsified acids,in: SixthInternational PetroleumTechnolo gyConference,March26–28, 2013 Beijing, China, 2013.
105. A. Rostami, H.A. Nasr-El-Din, Review and evaluation of corrosion inhibitors used in well stimulation, in: SPE International Symposium on Oilfield Chemistry, April 20–22, 2009, The Woodlands, TX, Society of Petroleum Engineers, 2009.
106. J. Yang, V. Jovancicevic, S. Mancuso, J. Mitchell, High performance batch treating corrosion inhibitor, in: Corrosion 2007 Conference & Expo, NACE International, March 11–15, 2007, Nashville, TN, 2007.
107. National Toxicology Program, Report on Carcinogens, 12th Ed., U.S. Department of Health and Human Services, Research Triangle Park, NC, 2011.
108. C. Sitz,W. Frenier, C.Vallejo,Acid corrosion inhibitors with improved environmental profiles, in: SPE International

Conference and Exhibition on Oilfield Corrosion, May 28–29, 2012, Aberdeen, UK, Society of Petroleum Engineers, 2012.

109. M. Crabtree, D. Eslinger, P. Fletcher, M. Miller, A. Johnson, G. King, Fighting scale—removal and prevention, Oilfield Rev. 11 (1999) 30–45.
110. D.R. Watkins, J.J. Clemens, J.C. Smith, S.N. Sharma, H.G. Edwards, Use of scale inhibitors in hydraulic fracture fluids to prevent scale build-up, Union Oil Company of California, in: U.S. Patent 5,224,543, 1993.
111. M.M. Brezinski, T.R. Gardner, W.M. Harms, J.L. Lane, K.L. King, Controlling iron in aqueous well fracturing fluids, Halliburton Energy Services, Inc, in: U.S. Patent 5,674,817, 1997.
112. K.C. Taylor, H.A. Nasr-El-Din, M.J. Al-Alawi, S. Aramco, Systematic study of iron control chemicals used during well stimulation, Soc. Pet. Eng. J. 4 (1999) 19–24.
113. W.R. Dill, G. Fredette, Iron control in the Appalachian Basin, in: Society of Petroleum Engineers Eastern Regional Meeting, November 9–11, 1983, Pittsburgh, PA, Society of Petroleum Engineers, 1983.
114. Z. Zhou, D.H.S. Law, Swelling Clays in Hydrocarbon Reservoirs: The Bad, the Less Bad, and the Useful 1998.057, Alberta Research Council, Edmonton, AB, 1998.
115. Z.J. Zhou, W.O. Gunter, R.G. Jonasson, Controlling formation damage using clay stabilizers: a review, in: SPE Annual Technical Meeting, June 7–9, 1995, Calgary, Canada, Society of Petroleum Engineers, 1995.
116. S.L. Berry, J.L. Boles, H.D. Brannon, B.B. Beall, Performance evaluation of ionic liquids as a clay stabilizer and shale inhibitor, in: Society of Petroleum Engineers International Symposium and Exhibition on Formation Damage Control, February 13–15, 2008, Lafayette, LA, Society of Petroleum Engineers, 2008.

117. E. Chevet, G.M. Lemaitre, I.D. Katinka, Low concentrations of tetramethylammonium chloride increase yield and specificity of PCR, Nucleic Acids Res. 23 (1995) 3343.
118. National Research Council, Dietary Reference Intakes for Thiamin, Riboflavin, Niacin, Vitamin B6, Folate, Vitamin B12, Pantothenic Acid, Biotin, and Choline, The National Academies Press, Washington, DC, 1998, in press.
119. Balchem Corporation, Choline Product for Clay Stabilization Brochure, Balchem Corporation, 2010, http://www.balchem.com/sites/default/files/ Choline%20Salts%20-%20Clay%20Stabilizer.pdf(accessed August 10, 2013).
120. R.T. Whalen, Viscoelastic surfactant fracturing fluids and a method for fracturing subterranean formations, in: U.S. Patent 6,035,936, 2000.
121. Organization for Economic Cooperation and Development (OECD), OECD Guidelines for the Testing of Chemicals, Section 2: Effects on Biotic Systems, Organization for Economic Cooperation and Development (OECD), 2013

Chapter 5

Chemical Constituents and Analytical Approaches for Hydraulic Fracturing Waters

Imma Ferrer and E. Michael Thurman

Department of Environmental Sustainability, University of Colorado, Boulder, CO, USA

ABSTRACT

Hydraulic fracturing fluids contain a mix of organic and inorganic additives in an aqueous media. The compositions of these mixtures vary according to the region or company use, thus making the process of identifying individual compounds difficult. The analytical characterization of such mixtures is important in order to understand the transport, environmental fate and ultimate potential health impact in various water compartments associated with hydraulic fracturing. Organic compound classes include solvents, gels, biocides, scale inhibitors, friction reducers, surfactants and other related compounds. These contaminants are usually present in trace amounts, so sophisticated analytical methodologies are needed in order to fully characterize the chemical composition of fracking fluids. The current state of knowledge of chemical components and approaches for their analysis is reviewed here. In recent years, modern analytical methodologies, such as gas chromatography–mass spectrometry (GC–MS) have been specifically used to identify organic chemical components of fracking fluids and/or flowback and produced waters associated with the process of hydraulic fracturing. Other techniques such as liquid chromatography–mass spectrometry (LC–MS) have not been explored in detail yet. In this review a detailed description of chemical constituents present in hydraulic fracturing waters will be given, as well as an evaluation of the analytical techniques used for their unequivocal determination.

INTRODUCTION

Hydraulic fracturing, commonly known as "fracking", is the process of extracting natural gas from shale formations by using a pressurized drilling technique. Fracturing fluids are injected into deep wells under high pressure conditions. This process allows fracturing of the geological formation, increasing permeability and extracting oil and gas. The latest advances in horizontal drilling (multiple wells drilled from one surface location) and improvements in fracturing techniques have made this technique very popular and economically competitive for natural resources recovery. In the U.S., the number of natural gas wells has increased by 200,000 in the last two decades, and it is projected to increase gas production to about 1065 billion m^3 $year^{-1}$ by 2040 [1]. The fast advances in this type of technique has allowed prices of natural gas to become lower and has made hydraulic fracturing a promising new energy extraction technology of this century. Recently, the process of hydraulic fracturing has been in the news for its controversial nature.

Some reports have tried to examine public perceptions on the topic [2]. A tendency for people to view this process as an environmental issue, rather than energy production, has been highlighted [3]. As a result new policies and regulations are coming in place to take into account potential environmental risks associated with these activities. For example, a recent study by Eaton [4] pointed out that some of the actual regulatory frameworks for hydraulic fracturing are not adequate to prevent contamination of water supplies in the city of New York. Another study proposed protective measures in field construction and maintenance in order to minimize exposure of hydraulic fracturing activities in waste streams that were found to contain exceeding limits of contaminants [5]. Therefore, all these facts and recent studies will help change some of the policies and management of such activities in order to protect the environment.

The process of hydraulic fracturing consists basically of 3 steps: (i) inject water, sand and chemicals (known as fracking fluids) down a horizontal drilled well to fracture the rock formation, (ii) extract the natural gas released from the formation through the well, and (iii) dispose or treat the produced water that was used to fracture the well. Therefore, one of the obvious immediate consequences from the fracturing process is the substantial volumes of water that return to the surface once the pressure is released. It is estimated that around 2–5 million gallons are used per well when performing hydraulic fracturing [6]. Because the water that is injected into the well (flowback water) gets mixed with the native geologic formation water (produced water) these two terms are often confused. Some papers have technically addressed the distinction between flowback water and produced water [7], [8] and [9]. Flowback is the immediate return of injected fluids and water and produced water is usually mixed with formation water native to the well. Other definitions distinguish these two types of water as the initial water that is recovered (for flowback) and the long-term water (for produced water) that comes back while the gas production is taking place [8] and [9]. Thus, there are two types of contaminants: those that are native in shale formations and those that are in the fracking fluids used to fracture the wells. Depending on which water is investigated, the fraction of one or another type of contaminants might vary. One study [9] reported that the levels of some hydraulic fracturing chemicals decreased rapidly over the first 20 days of water recovery while others can remain up to 250 days. The water that returns to the surface requires treatment before being disposed into the environment or being re-used. In many cases, the disposal is through deep injection into water wells, thus potentially affecting groundwater resources. In these cases, the presence of fracking fluid components makes these waters warrant to be investigated for their potential environmental fate in water disposal or water re-use.

One of the major concerns is that the specific chemicals used in fracking fluids could potentially contaminate ground and surface water supplies. This concern is even more enhanced if we take into account that oil and gas companies are reluctant to share the details of what's in their proprietary fluid mixes. Recent regulations in some states require the companies to disclose what is in their mixtures [10]. However, the listed ingredients are usually related to broad chemical categories and do not reflect the individual exact compositions of the mixtures. An example of this is the FracFocus Chemical Disclosure Registry [11], which contains broad and general information of chemicals used at different wells across the U.S. Scarce information has been published in the literature on specific determination of hydraulic fluid components. In order to understand the environmental fate of many of these compounds, it is necessary to be able to detect them in their sources (fracking fluid and flowback/produced waters), as well as to perform a rigorous monitoring through the various processes established in each hydraulic fracturing operation (i.e., treatment of water, water disposal). For this reason, advanced and modern analytical techniques are needed in order to fully characterize specific individual components of fracking fluids. In this review we focus on which analytical techniques and strategies have been used or could be potentially used for the detection and identification of the majority of the organic chemicals associated with hydraulic fracturing operations.

ENVIRONMENTAL IMPACT OF HYDRAULIC FRACTURING

When reviewing the environmental impact of hydraulic fracturing there are two main areas to consider. One is the potential atmospheric pollution that occurs on or near the fracturing sites. A recent study [12] focused on how the production of oil and natural gas affected air quality by using data from contaminants such as nitrogen oxides (NO_x), methane, ozone, volatile organic compounds (VOC), and other related pollutants. Similarly, Roy et al. [13] described an air emissions inventory, including estimates for individual emissions of NO_x and VOCs, in the Marcellus Shale region. The second important area is the impact on water sources. The most comprehensive study showing the impact of hydraulic fracturing in water quality was recently published by Vidic et al. [14]. Several papers have addressed the specific impacts in water resources as well as the slow contamination of shallow groundwater [15], [16], [17], [18], [19], [20],[21] and [22]. Most particularly some studies have identified certain pathways of water contamination, which consist mainly of the following: transportation spills, stray gas contamination, leaks from well casing, leaks from fractured rocks, site discharges, accumulation of radioactive elements in soil and sediments, and ultimately wastewater disposal [14], [15] and [19]. For example, methane has been shown to migrate from the shale formation to groundwater in several studies [17], [20], [21], [23] and [24]. Thermogenic methane sources have been associated with hydraulic fracturing operations whereas biogenic methane sources are typical of non-active sites [20]. Likewise, total dissolved solids including cations, metals and radioactive elements have been detected in surface and groundwater sources [17], [22], [25] and [26]. Interestingly, organic compounds have not been studied as often as gases or inorganic elements. Synthetic and volatile organic compounds have been analyzed in hydraulic fracturing waters [25]. Strong et al. [27] identified several organic compounds in fracking fluids and produced waters. Similarly, Orem et al.[9] and [28] published an extensive study on the analysis of a number of organic compounds originating from formation waters and produced waters. Thurman et al. [29] developed a methodology using a database for the identification of a subgroup of ethoxylated surfactants and proposed that this group of compounds could be potentially be used as tracers in surface and groundwaters associated with hydraulic fracturing operations. Thus, there is a current need to develop and identify specific methods that can detect the occurrence of chemical additives present in hydraulic fracturing fluids, so they can be used as tracers of water contamination associated with these processes.

Another environmental issue associated with flowback waters is the treatment of the large volumes of water used in hydraulic fracturing, rather than deep waste disposal. A recent study estimates the life cycle water consumption in a shale gas well and proposes four different scenarios for the wastewater generated[16]. Similarly, a study by Rahm et al. [30] examines several wastewater management trends seen in the last few years and suggests practices for water management. A recent work examined the possibility of conventional biological treatment of hydraulic fracturing waters [31]. But again, in all these cases it is important to first characterize the potential contaminants present in hydraulic fracturing waters in order to follow and monitor specific compounds.

CHEMICAL CONSTITUENTS

If one considers both, flowback water and produced water then a distinction between different types of constituents or contaminants needs to be made. For flowback waters, which usually refer to water that it is immediately returned from the well after the pressure is released, the focus should be on chemicals or additives that were injected with the fracking fluid. For produced waters, which come from the long-term mixing within the formation, one should consider both fracking additives and constituents native to the formation. These constituents usually comprise among others: natural occurring radioactive elements, heavy metals, hydrocarbons, and volatile and semi-volatile organic compounds. In Sections 3.1 and 3.2, a detailed description for each group of contaminants is given.

Native Constituents from the Geologic Formation

Produced water is mainly composed from formation water (naturally occurring water in the geologic shale). Therefore, the natural components present in this type of formation will be present when the water is recovered from a well for a long-term period. These substances can originate directly from the formation water, from the formation rock present in shale and from oil or gas present in the formation. Some of the most important classes of constituents are described as follows:

- Naturally occurring radioactive materials (uranium, thorium, radium, radon, strontium and potassium). Barbot et al. [26] found radium and uranium concentrations in Marcellus Shale produced water at thepCi/L level. Similarly, another study found significant levels of radium in flowback water from the same geographical area [25]. In other cases, the strontium isotope ratios have been used as indicators to determine if produced waters from shale formations have an impact in other sources of water such as surface or groundwaters [32]. Radium has also been identified in produced waters from the Northern Appalachian Basin [33]. Warner et al. [24] detected barium and radium levels in produced waters but the concentrations were substantially reduced in treated effluents. However, ^{226}Ra levels in stream sediments were 200 times higher than upstream and exceeded radioactive waste disposal threshold regulations.
- Inorganic substances and metals (aluminum, arsenic, barium, bromine, cadmium, chloride, chromium, iron, manganese, mercury, nickel, sodium, vanadium and zinc). Because of the nature of formation water, the salinity of these type of

samples is always high, thus being Br, Cl and Na the most regularly detected elements. For example, the most common elements found in shale produced water from the Marcellus Shale (PA, USA) were Ba, Br, Ca, Cl, Na, and Sr [24], [26] and [32]. Abualfaraj et al. [25] reported exceeding maximum concentration levels (MCL) for drinking water standards for Al, Cl, Fe and Mn in flowback water samples.

- Volatile organic gases ($CH_4$and CO_2). The release of these gases from processing plants has been documented in several states of the U.S. and the potential for air contamination is well-known [34]. Osborn et al. [20] found concentrations of methane up to 64 mg/L in drinking water sources in upstate New York. The authors also correlated the detected ^{13}C isotope composition of methane with thermogenic sources and hydraulic fracturing locations. Methane was found in 82% of drinking water samples (n = 141) with higher concentrations in locations that were <1 km from natural gas wells [23]. Ethane and propane were also detected in the same study. In another study, CH_4 was found in 63% of the samples analyzed from 127 drinking water wells [22]. Similarly, other authors have found a strong correlation between the presence of CH_4 and proximity to fracturing sites [21]. In all these studies, the isotopic composition of methane was correlated with thermogenic processes and thus with hydraulic fracturing operations. Stable isotope analysis of C_1–C_4 gases was used by Sherwood et al. [35] to identify sources of natural and produced hydrocarbon gases and to develop fingerprinting protocols that permit tracing of gases to specific geological formations.
- Hydrocarbons (polycyclic aromatic hydrocarbons (PAHs), heterocyclic compounds, phenols, long chain fatty acids, alkyl benzenes and aliphatic hydrocarbons). Migration of shale-derived hydrocarbons into the formation water is also possible during hydraulic fracturing processes. For example the total concentration found for PAHs in produced water from coal bed methane formations ranged up to 23 μg/L

and 100 μg/L in the Powder River Basin (WY, USA) [28] and the Black Warrior Basin (AL, USA) [9]. Low molecular weight (2-ring) PAHs were the abundant compounds found in most of the locations studied. Nonylphenols were also found in the same study at concentrations up to 7.9 μg/L. On the other hand, the authors found that the most common hydrocarbons found in gas shale formations were more aliphatic rather than aromatic. For example, long chain fatty acids were detected in produced water from gas shale deposits. The authors suggested that these compounds were biodegradation products of geopolymeric substances in the shale, as shown in a previous study[36]. Orem et al. [28] showed that hydrocarbons recovered from produced water from a coal bed methane well decreased over a one-year period. This phenomena was mainly due to the migration of water from outside the coal aquifer once the water that was associated with the coal for long periods was totally removed. This causes the dissolved organic compounds concentration to decrease over time. Contrarily, in a gas shale well, the concentration for hydrocarbons decreased rapidly after 20 days of production [9] due to the influence of hydraulic fracturing fluids. However in these cases, there is a residual concentration of organic compounds found even after 250 days, probably the result of mixing of flowback waters with produced waters.

Chemical Composition of Fracking Fluids

Chemicals used for hydraulic fracturing can be divided into several groups. One important group is the drilling fluids which are used to lubricate and are used for the injection of water down the well. Another group is the chemicals used to cement the steel pipes or casing to the sides of the well. A third group is a more general group including chemicals that are used as corrosion inhibitors, biocides and emulsifiers. All this complexity is even more enhanced if we take into account that each well may be fractured differently and the geology of the specific site might require a different chemistry. As an example, between 2005 and 2009, 14 different oil and gas service companies used more than 2500 hydraulic fracturing products containing 750 chemicals and other components in a total of 94 million gallons. The list of the 750 chemicals and other components used in hydraulic fracturing products between these years can be found in [37]. Some of these components, such as methanol or aromatic hydrocarbons are regulated contaminants under the Safe Drinking Water Act and hazardous air pollutants under the Clean Air Act. This is the reason why in the last few years an effort to use less toxic and more amenable chemicals has been made. In fact, in some cases the U.S. EPA has already worked with major gas producers to eliminate some of the toxic additives used in fracking fluid [37].

Generally the individual chemical compounds used in fracturing fluids depend on the company involved and site specific characteristics of the formation. Table 1 shows the representative compounds of a hydraulic fracturing fluid and their purposes. The physical and chemical characterization of all these compounds was recently reviewed by Stringfellow et al. in a very comprehensive study [7] in which the authors used public available chemical information databases. In this study, 81 common hydraulic fracturing additives were categorized, from these 55 of the compounds were organic and 27 were considered biodegradable. Only 3 compounds were classified as oral toxins. The study was focused on the evaluation of these chemicals for potential environmental risk and as a guideline to assess the fate of specific compounds into the environment.

Table 1: Chemical content of a fracturing fluid and specific purposes for hydraulic fracturing operations

Additives	**Chemical composition**	**Purpose**	%
Water	H_2O	Main carrier (base carrier fluid)	90.800
Sand	Water and crystalline silica (quartz)	Propping agent use to hold open fractures	8.500

Acids	Hydrochloric acid, acetic acid,	Clean and help dissolve minerals and initiate cracks in the rock	0.150
Clay stabilizers	Choline chloride, tetramethylammonium chloride	Prevent swelling of clays found in shale	0.120
Scale inhibitors	Carboxylic acids and acrylic acid polymers	Prevent formation of scale (mineral) deposits in the pipe	0.090
Surfactants	Amido-amines, quaternary amines, phosphate esters, alcohol polyethoxylates, ethylenene glycols, isopropanol	Increase the viscosity of the fluid	0.075
Friction reducers	Polyacrylamide	Minimize friction between the fluid and the pipe, thus allowing to pump at a higher rate	0.070
Breakers	NaCl and KCl	Reverse crosslinking allowing the production gas to flow through	0.060
Biocides	Gluteraldehyde, DBNPA, quaternary ammonium compounds	Prevent bacteria growth in water	0.060
Gels	Guar gum	Thicken the water to suspend the sand and also increases viscosity of the fluid to deliver proppant more efficiently	0.050
pH adjusting agents	Sodium or potassium carbonate	Maintain effectiveness of crosslinkers	0.010
Crosslinkers	Borate salts	Maintain fluid viscosity as temperature increases	0.007
Iron control	Citric acid	Prevent precipitation of iron oxides	0.006
Corrosion inhibitors	Amines, amides and amido-amines	Prevent corrossion of the pipe	0.002

It is important to note that the chemical additives used in hydraulic fracturing fluids account for only up to 0.5–1% of the total fluid used (Fig. 1). However, these are the only unique compounds that can be used as tracers in the environment as discussed above; therefore their detailed description and available techniques for their identification in water samples is highly important, and for these reasons they have been the main objective of this review, which are now discussed in conjunction with Table 1.

a. Gels. These classes of chemicals are used to increase viscosity of the fracking fluid, thus allowing an enhanced sand suspension in the mix and ultimately delivering the proppant (sand) into the geologic fractures. They are linear or cross-linked polymers consisting of several monomer groups. The most common gelling agent is guar gum and its carboxylated derivatives [7]. Other gelling agents are based on cellulose. These compounds are usually found also in cosmetics and food products. Guar gum is a polysaccharide composed of galactose and mannose. The backbone is a linear chain of -1,4-linked mannose monomers to which alpha-1,6-galactose units are linked, forming short side branches. Guar is considered to be biodegradable [31], thus not being a potential toxic environmental contaminant. Gel stabilizers such as organic solvents (methanol, ethanol, isopropanol and ethylene glycol) are also used in fracking fluids to prevent decomposition [11]. The fluids that contain gelling agents are commonly known as "gel-frac".
b. Crosslinkers. These compounds maintain fluid viscosity by chemically binding individual gel polymer molecules, thus allowing a better transport of the proppant into the well. Commonly used crosslinkers are: borate salts, inorganic complexes with zirconium or aluminum, monoethanolamine, monoethylamine, ammonium chloride, ethylene glycol and potassium hydroxide.
c. Friction reducers. These additives are used to reduce interfacial tension between the fluid and the surface of the pipe, as well as to maintain laminar flow while pumping. Fracturing fluids containing this type of additive are commonly known as "slick water". The main active ingredient used is usually polyacrylamide dispersed in a hydrocarbon carrier.

Sometimes these additives are used instead of gels and have proven to be more effective by reducing fracture plugging [38].

d. Breakers. These compounds usually get introduced down the well after fracturing has occurred. These breakers reverse the crosslinking, thus reducing the viscosity and allowing the production gas to flow through, increasing productivity. These additives are also helpful when recovering fracturing fluids from a well, implying that they will be present in flow-back and produced waters. Sometimes these additives are included in gels in encapsulated form, in order to prevent premature breaking of the polymer. There are two types of breakers: enzyme breakers and inorganic breakers. Enzyme breakers are mainly proteins (hemicellulases) that degrade cellulose polymers into smaller sugars. At temperatures below 66 °C enzyme breakers work well to degrade guar gels, but at temperatures above 95 °C they are not so effective and then inorganic breakers are recommended [39]. Inorganic breakers can be common salts such as calcium chloride, sodium or potassium chloride and ammonium sulfate [11].
e. pH adjusters. Acids or bases can be added to fracturing fluids to increase the effectiveness of the polymers and crosslinkers being used. Depending on which type of polymer or crosslinker is used a high pH or a low pH may be necessary. Some examples of pH adjusters are: acetic acid, sodium or potassium hydroxide, sodium or potassium carbonate.
f. Acids. They are mainly used to clean the wellbore and to dissolve the minerals present in the geological formation, prior to the injection of hydraulic fluids. By dissolving the minerals, additional flow paths for gas or oil are opened up and can make their way to the wellbore, thus increasing production of the well. The most common acid for this purpose is hydrochloric acid (15% active), which can dissolve calcite ($CaCO_3$), a mineral generally found in shales.
g. Corrosion inhibitors. These compounds are used to prevent potential corrosion in the casing itself caused by the use of

acids and salts. They are usually added to the acids themselves; they form a protective layer on metal well casings. Chemically, they consist of acetaldehyde, acetone, formic acid, thiourea, amines, amides or amido-amines. So, a whole variety of chemical compounds can be used as corrosion inhibitors.

h. Scale inhibitors. Mineral solids present in formation waters can precipitate during the production stage when getting in contact with injection fluids. For this reason, scale inhibitor compounds are used in the same fracturing fluids. The most common chemistries for these types of compounds are carboxylic acids, acrylic acid polymers and phosphonic acid salts.
i. Iron control. As mentioned above hydrochloric acid is usually added to dissolve minerals in the formation, but it could also dissolve any iron oxide from the casing as well and therefore there could be a potential precipitation. Moreover, some shale formations are high in iron content. This would block the flow channels in the well, so for this reason an iron control agent is added to fracturing fluids. The most common iron control compounds are citric acid, acetic acid, thioglycolic acid and EDTA. These compounds complex the dissolved iron, thus preventing it from precipitating in the wellbore.
j. Clay stabilizers. They are used to prevent the swelling of clays found in shale formations. Swelling of clays could reduce permeability by up to 90%, thus reducing well productivity [40]. Clay stabilizers are usually ion exchange compounds such as choline chloride, tetramethylammonium chloride, potassium and sodium chloride. Quaternary amines have been also used in fracturing fluids by coating the clay particles present in the shale, and thus preventing the adsorption of water [41].
k. Biocides. The water used in hydraulic fracturing fluids often contain a number of bacteria that could cause potential problems during the extraction process, such as clogging of the wellbore, corrosion of the metal equipment used

or by even degrading the chemicals used in the fluids. For these reasons, biocides are added to the fracturing fluids to sanitize and reduce the concentration of bacteria present in them. Biocides is the group with most variety of different chemistries used. Historically, the compound most generally used has been glutaraldehyde and their derivates. However this compound is being replaced by other non-toxic or green alternative biocides. Other compounds used as biocides are tetrakis hydroxymethyl phosphonium sulfate, 2-2-dibromo,3-nitrilopropionamide (DBNPA), quaternary ammonium compounds (i.e., dodecyl dimethyl ammonium chloride and alkyl dimethyl benzyl ammonium chloride) and common sodium hypochlorite or bleach. All biocide usage is regulated by the Federal Insecticide, Fungicide and Rodenticide Act and every product must be registered by EPA [41].

I. Surfactants. These types of compounds are used to reduce interfacial tension between the hydraulic fluids and the shale. They are also used to remove possible emulsions formed by the mix of oil and water in the well. This is probably the broadest class of chemical compounds and the most complex group of additives used in hydraulic fracturing fluids. This is due to the fact that sometimes surfactants can also be used as gelling agents, crosslinkers, corrosion inhibitors or even biocides; thus, they fall into various categories and their intrinsic chemical properties make this class of compounds highly complex to study. They can be divided mainly into 4 group classes: non-ionic, anionic, cationic and amphoteric compounds. The most common class used to date have been lauryl sulfates, but in the last few years new classes of surfactants have been considered and will be discussed later. Due to the lack of individual CAS numbers and specific chemical composition, this class of compounds has been always omitted in the hydraulic fracturing papers that have been found thus far in the literature. Only recently, an effort was made to analyze and characterize the individual chemical composition of a class of surfactants found in hydraulic fracturing waters [29].

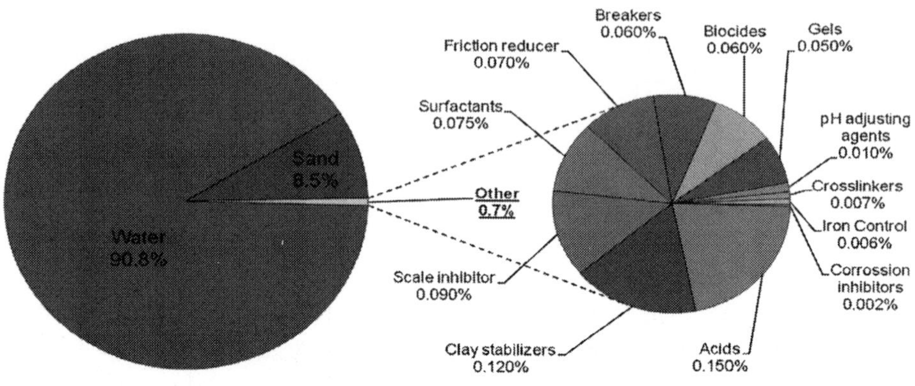

Figure 1: Chemical constituents of a hydraulic fracturing fluid.

ANALYTICAL METHODOLOGIES

Since hydraulic fracturing waters may contain a whole different variety of chemicals (both natural from the formation shale and additives added to the fluids) as seen in previous sections, the diversity of analytical techniques used for their identification will be broad as well. We have distinguished the different types of sample processing and detection techniques for each group of chemical constituents described in this paper (see Fig. 2). In the next subsections a detailed description of each of the analytical approaches used will be given, and some examples of referenced works using such methodologies will be discussed.

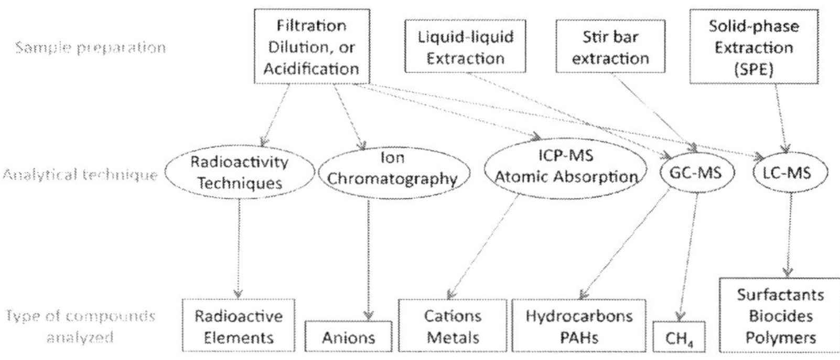

Figure 2: Analytical approaches for the analysis of chemical constituents of hydraulic fracturing waters.

Sample Preparation Techniques

Depending on the analyte to be determined and which analytical technique is used, sample preparation can vary. A comprehensive study describing how hydraulic fracturing waters (fluids and flowback and produced waters) are processed does not exist in the literature. Most of the studies for flowback and produced water analyzed the samples directly, applying only filtration (through 0.45 μm filters) as a pre-treatment procedure [9], [22], [24], [26] and [32]. Usually, flowback water samples have a high TOC, and for this reason, no further treatment is needed in order to detect inorganic and organic compounds. However, produced water samples are much more diluted than flowback water when considering organic compounds, and for this reason, some type of pre-treatment may be necessary. For example, a liquid–liquid extraction procedure using dichloromethane has been applied for extraction of hydrocarbons from produced waters prior to GC–MS analyses [9]. Another procedure involving extraction with a stir bar after dilution of the samples has been described elsewhere [27]. No studies on solid-phase extraction (SPE) of flowback or produced waters have been reported to date. Fig. 3 shows the comparison between a flowback and a produced water sample analyzed by

LC-TOF–MS. As it can be seen in this figure, organic compounds in flowback water can be clearly distinguished from the background signal, whereas produced water samples exhibit a lower total ion current, thus making more difficult the identification of organic compounds. It appears from this review that SPE techniques will have to be developed in the near future in order to better characterize produced water samples by pre-concentrating trace components present in them.

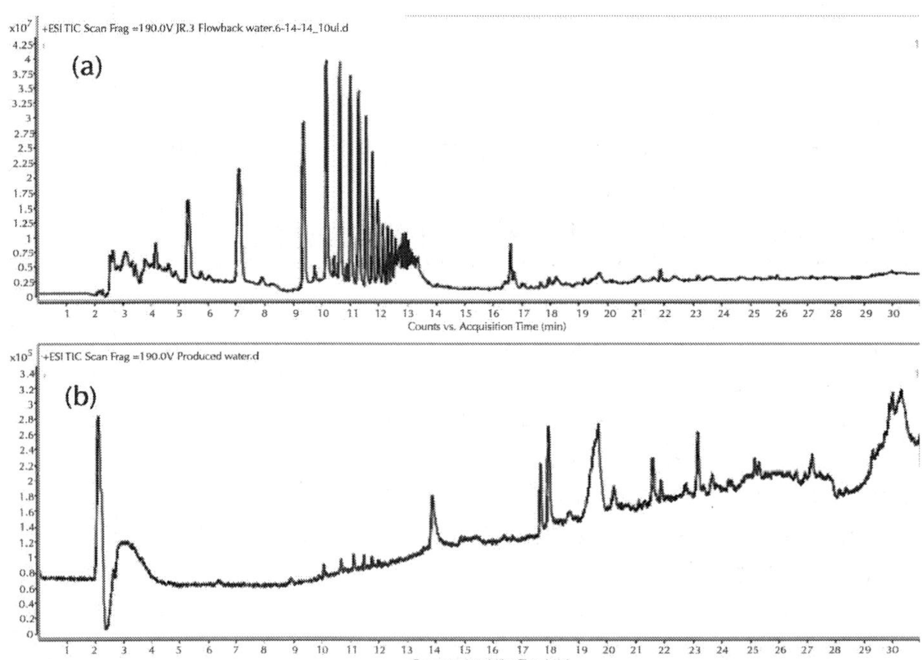

Figure 3: LC/Q-TOF–MS analysis of a (a) flowback and (b) produced water.

Inorganic Analyses Techniques

Inductive coupled plasma mass spectrometry (ICP-MS) has been widely used for the identification of inorganic elements and metals such as Na, Ca, Ba, Fe, Mg, Sr [27] and [32], although atomic absorption has also been used for the same purpose in other works [26]. Ion chromatography is the technique of choice for the identification of major anions present in produced waters [24]. Strontium isotopes are usually analyzed using a thermal ionization mass spectrometer [24] and [32]. In all these cases, sample preparation is minimal, requiring only a filtration and/or dilution of the sample, as well as addition of acids or other reagents for the analysis of the metals.

Organic Analyses Techniques

Organic compounds in flowback or produced waters have not been widely identified. Hydrocarbons (present also in formation waters) are probably the class of compounds that has been studied in greatest detail. Only a few papers, mainly in the last decade, have reported the identification of organic compounds in flowback and produced waters [27] and [28]. Furthermore, few attempts to identify organic additives present in hydraulic fluids have been made. Orem et al. [9] reported the identification of a heterocyclic compound used as a biocide and tridecane (additive) in produced water. Strong et al. [27] reported the identification of several groups of aromatic, aliphatic, cycloaliphatic and PAH compounds, although no individual elemental compositions were made. A recent work by our group [29] reported the presence of polyethylene glycols and alkyl ethoxylates, used as surfactants, in flowback waters, and individual chemical compositions were assigned for each identified compound. In the next subsections, an overview of the instrumental approaches used for some of the organic compounds present in hydraulic fracturing waters is given.

GC–MS

Gas chromatography–mass spectrometry is best suited for the analysis of volatile and polar compounds. The most common use of this technique is perhaps the analysis of methane in flowback and produced waters. Methane concentrations and its isotopic composition is measured after chromatographic separation followed by combustion and dual-inlet isotope ratio mass spectrometer [20], [21] and [35]. By measuring ^{13}C isotope ratios, the source of methane gases (biogenic or thermogenic) can be established.

Other organic compounds such as PAHs, aromatic amines, heterocyclic compounds, phthalates and phenols were analyzed using a GC–MS instrument [9] and [28] operated in scan mode. Time-of-flight detectors have also been used in combination with GC–MS for the analysis of the same type of compounds in flowback and produced waters [27]. With TOF detectors, a more detailed and exact composition of each individual compound can be obtained, which aids in the complex information that it is obtained after the analysis of these type of samples [42]. But these techniques still lack some identification power. As stated by Strong et al. [27], around 25% of the compounds detected in a single GC–GC-TOF–MS analysis can not be identified to the level of a single chemical structure. Many of these compounds are isomeric and identification requires good mass accuracy and resolving power as well as existence of broad chemical databases for unequivocal identification.

LC–MS

Surprisingly no methodological analytical techniques based on liquid chromatography–mass spectrometry have been described in the literature for the analysis of hydraulic fracturing waters, with one exception. That exception is our own work published in 2014 [29]. In this work, an LC–Q-TOF–MS using high-resolution was used for the identification of a sub-class of ethoxylated surfactants (alcohol polyethoxylates and polyethylene glycols) in flowback and produced waters from hydraulic fracturing wells. A unique combination of TOF and the Kendrick mass defect was applied to identify the distinctive chemical composition of this class of surfactants seen in Fig. 3(a).

LC–MS is a suitable technique for both polar and hydrophobic compounds used in hydraulic fracturing operations such as biocides, corrosion and scale inhibitors, gels and surfactants. Our prediction for the next few years is that we will see a number of other methods employing LC–MS techniques for the identification of other chemical additives in hydraulic fracturing fluids and waters. Specifically, time-of-flight will be used for the elemental composition identification of individual compounds and triple quadrupole methods using tandem mass spectrometry will be used to accurately follow and monitor these compounds in associated waters from hydraulic fracturing operations (i.e., treatment technologies and monitoring at low levels in the environment). Similarly to what has happened for the analysis of small organic molecules (pesticides and pharmaceuticals) will happen as well in this field of research. The reason accurate mass will be favored is the fact that unknowns can be identified, even without authentic standards when using MS–MS analysis. This is highly important given the difficulty of obtaining standards of chemicals additives used in hydraulic fracturing fluids.

CONCLUSIONS

Hydraulic fracturing using horizontal drilling is a relatively new industrial process aimed to the extraction of natural gas from a shale formation. High volumes of water, sand and additives are needed in order to create fractures in a given well. After extraction, waters associated with the process (called flowback and produced waters) have to be properly disposed. For this reason, a full characterization of these types of waters is necessary, as well as an understanding of environmental transport and fate of potential associated contaminants. Therefore, understanding what components need to be monitored in the environment is the key to this problem. An overview of the chemical additives used for hydraulic fracturing is presented here. Available data on the identification of both natural components from a geological formation and chemical additives in flowback and produced waters is reviewed. Analytical techniques that can detect these types of compounds are also presented and discussed in this paper.

ACKNOWLEDGMENTS

We would like to thank Dr. James Rosenblum for providing the JR-3 sample.

REFERENCES

1. Annual Energy Outlook 2014 with projections to 2040, in: U.S. Energy Informations Administration (EIA) (Ed.), DOE/EIA-0383(2014), U.S. Energy Informations Administration (EIA), Washington, DC, 2014.
2. H. Boudet, C. Clarke, D. Bugden, E. Maibach, C. Roser-Renouf, A. Leiserowitz, Fracking controversy and communication: using national survey data to understand public perceptions of hydraulic fracturing, Energy Policy 65 (2014) 57–67.
3. C. Davis, J.M. Fisk, Energy abundance or environmental worries: analyzing public support for fracking in the United States, Rev. Policy Res. 31 (2014) 1–16.
4. T.T. Eaton, Science-based decision-making on complex issues: Marcellus Shale gas hydrofracking and New York City water supply, Sci. Total Environ. 461 (2013) 158–169.
5. P.F. Ziemkiewicz, J.D. Quaranta, A. Darnell, R. Wise, Exposure pathways related to shale gas development and procedures for reducing environmental and public risk, J. Nat. Gas Sci. Eng. 16 (2014) 77–84.
6. G.E. King, Hydraulic fracturing 101: what every representative, environmentalist, regulator, reporter, investor, university researcher, neighbor andengineer should know about estimating frac risk and improving frac per-formance in unconventional gas and oil wells, SPE Hydraulic Fracturing Technology Conference February 6–8, 2012, The Woodlands, TX, Society of Petroleum Engineers, 2012.
7. W.T. Stringfellow, J.K. Domen, M.K. Camarillo, W.L. Sandelin, S. Borglin, Physical chemical, and biological characteristics

of compounds used in hydraulic fracturing, J. Hazard. Mater. 275 (2014) 37–54.

8. J.A. Veil, C.E. Clark, Produced-water-volume estimates and management practices, SPE Prod. Oper. 26 (2011) 234–239.
9. W. Orem, C. Tatu, M. Varonka, H. Lerch, A. Bates, M. Engle, L. Crosby, J. McIntosh, Organic substances in produced and formation water from unconventional natural gas extraction in coal and shale, Int. J. Coal Geol. 126 (2014) 20–31.
10. [10] A.L. Maule, C.M. Makey, E.B. Benson, I.J. Burrows, M.K. Scammell, Disclosure of hydraulic fracturing fluid chemical additives: analysis of regulations, New Solut. 23 (2013) 167–187.
11. FracFocus, FracFocus Chemical Disclosure Registry, Ground Water Pro-tection Council and Interstate Oil and Gas Compact Commission, (2013) http:// fracfocus.org (accessed 12.14).
12. R.A. Field, J. Soltis, S. Murphy, Air quality concerns of unconventional oil and natural gas production, Environ. Sci. Processes Impact 16 (2014) 954–969.
13. A.A. Roy, P.J. Adams, A.L. Robinson, Air pollutant emissions from the development production, and processing of Marcellus Shale natural gas, J. Air Waste Manage. 64 (2014) 19–37.
14. R.D. Vidic, S.L. Brantley, J.M. Vandenbossche, D. Yoxtheimer, J.D. Abad, Impact of Shale gas development on regional water quality, Science 340 (2013) .
15. A. Vengosh, R.B. Jackson, N. Warner, T.H. Darrah, A. Kondash, A critical review of the risks to water resources from unconventional shale gas development and hydraulic fracturing in the United States, Environ. Sci. Technol. 48 (2014) 8334–8348.
16. M. Jiang, C.T. Hendrickson, J.M. VanBriesen, Life cycle water consumption and wastewater generation impacts of a Marcellus Shale gas well, Environ. Sci. Technol. 48 (2014) 1911–1920.

17. S.L. Brantley, D. Yoxtheimer, S. Arjmand, P. Grieve, R. Vidic, J. Pollak, G.T. Llewellyn, J. Abad, C. Simon, Water resource impacts during unconventional shale gas development: the Pennsylvania experience, Int. J. Coal Geol. 126 (2014) 140–156.
18. C. Gassiat, T. Gleeson, R. Lefebvre, J. McKenzie, Hydraulic fracturing in faulted sedimentary basins: numerical simulation of potential contamination of shallow aquifers over long time scales, Water Resour. Res. 49 (2013) 8310– 8327.
19. D.J. Rozell, S.J. Reaven, Water pollution risk associated with natural gas extraction from the Marcellus Shale, Risk Anal. 32 (2012) 1382–1393.
20. S.G. Osborn, A. Vengosh, N.R. Warner, R.B. Jackson, Methane contamination of drinking water accompanying gas-well drilling and hydraulic fracturing, Proc. Natl. Acad. Sci. U. S. A. 108 (2011) 8172–8176.
21. T.H. Darrah, A. Vengosh, R.B. Jackson, N.R. Warner, R.J. Poreda, Noble gases identify the mechanisms of fugitive gas contamination in drinking-water wells overlying the Marcellus and Barnett Shales, Proc. Natl. Acad. Sci. U. S. A. 111 (2014) 14076–14081.
22. N.R. Warner, T.M. Kresse, P.D. Hays, A. Down, J.D. Karr, R.B. Jackson, A. Vengosh, Geochemical and isotopic variations in shallow groundwater in areas of the Fayetteville Shale development, north-central Arkansas, Appl. Geochem. 35 (2013) 207–220.
23. R.B. Jackson, A. Vengosh, T.H. Darrah, N.R. Warner, A. Down, R.J. Poreda, S.G. Osborn, K.G. Zhao, J.D. Karr, Increased stray gas abundance in a subset of drinking water wells near Marcellus shale gas extraction, Proc. Natl. Acad. Sci. U. S. A. 110 (2013) 11250–11255.
24. N.R. Warner, C.A. Christie, R.B. Jackson, A. Vengosh, Impacts of Shale gas wastewater disposal on water quality in Western Pennsylvania, Environ. Sci. Technol. 47 (2013) 11849–11857.

25. N. Abualfaraj, P.L. Gurian, M.S. Olson, Characterization of Marcellus Shale flowback water, Environ. Eng. Sci. 31 (2014) 514–524.

26. E. Barbot, N.S. Vidic, K.B. Gregory, R.D. Vidic, Spatial and temporal correlation of water quality parameters of produced waters from devonian-age shale following hydraulic fracturing, Environ. Sci. Technol. 47 (2013) 2562–2569.

27. L.C. Strong, T. Gould, L. Kasinkas, M.J. Sadowsky, A. Aksan, L.P. Wackett, Biodegradation in produced waters: chemistry microbiology, and engineering, Abstr. Pap. Am. Chem. Soc. 246 (2013) .

28. W.H. Orem, C.A. Tatu, H.E. Lerch, C.A. Rice, T.T. Bartos, A.L. Bates, S. Tewalt, M.D. Corum, Organic compounds in produced waters from coalbed natural gas wells in the Powder River Basin, Wyoming, USA, Appl. Geochem. 22 (2007) 2240–2256.

29. E.M. Thurman, I. Ferrer, J. Blotevogel, T. Borch, Analysis of hydraulic fracturing flowback and produced waters using accurate mass: identification of ethoxylated surfactants, Anal. Chem. 86 (2014) 9653–9661.

30. B.G. Rahm, J.T. Bates, L.R. Bertoia, A.E. Galford, D.A. Yoxtheimer, S.J. Riha, Wastewater management and Marcellus Shale gas development: trends drivers, and planning implications, J. Environ. Manage. 120 (2013) 105–113.

31. Y. Lester, T. Yacob, I. Morrisey, K.G. Linden, Can we treat hydraulic fracturing flowback with a conventional biological process? The case of guar gum, Environ. Sci. Technol. Lett. 1 (2013) 133–136.

32. E.C. Chapman, R.C. Capo, B.W. Stewart, C.S. Kirby, R.W. Hammack, K.T. Schroeder, H.M. Edenborn, Geochemical and strontium isotope characterization of produced waters from Marcellus Shale natural gas extraction, Environ. Sci. Technol. 46 (2012) 3545–3553.

33. E.L. Rowan, M.A. Engle, C.S. Kirby, T.F. Kraemer, Radium content of oil- and gas- field produced waters in the northern

Appalachian Basin (USA)—summary and discussion of data, U.S. Geological Survey Scientific Investigations Report 2011–5135, 2011, pp. 31.

34. D.M. Kargbo, R.G. Wilhelm, D.J. Campbell, Natural gas plays in the Marcellus Shale: challenges and potential opportunities, Environ. Sci. Technol. 44 (2010) 5679–5684.
35. O.A.T. Sherwood, P.D. Travers, M.P. Dolan, Compound-specific stable isotope analysis of natural and produced hydrocarbons gases surrounding oil and gas operations, Comprehensive Analytical Chemistry Series: Advanced Techniques in Gas Chromatography-Mass Spectrometry (GC–MS–MS and GC-TOF–MS) for Environmental Chemistry, 61, 2014, pp. 347–369.
36. W.H. Orem, M.A. Voytek, E.J. Jones, H.E. Lerch, A.L. Bates, M.D. Corum, P.D. Warwick, A.C. Clark, Organic intermediates in the anaerobic biodegradation of coal to methane under laboratory conditions, Org. Geochem. 41 (2010) 997– 1000.
37. U.S. Environmental Protection Agency (U.S. EPA), Chapter 4: hydraulic fracturing fluids, Evaluation of Impacts to Underground Sources of DrinkingWater by Hydraulic Fracturing of Coalbed Methane Reservoirs, EPA 816-R-04-003, U. S. Environmental Protection Agency (U.S. EPA), Washington, DC, 2004.
38. J.Y. Wang, S.A. Holditch, D.A. McVay, Effect of gel damage on fracture fluid cleanup and long-term recovery in tight gas reservoirs, J. Nat. Gas Sci. Eng. 9 (2012) 108–118.
39. B. Zhang, A. Huston, L. Whipple, H. Urbina, K. Barrett, M. Wall, R. Hutchins, A. Mirakyan, A superior: high-performance enzyme for breaking borate crosslinked fracturing fluids under extreme well conditions, SPE Prod. Oper. 28 (2013) 210–216.
40. Z. Zhou, D.H.S. Law, Swelling Clays in Hydrocarbon Reservoirs: The Bad the Less Bad and the Useful, Alberta Research Council, Edmonton, AB, 1998.

41. R. McCurdy, High rate hydraulic fracturing additives in non-Marcellus unconventional shale, Proceedings of the Technical Workshops for the Hydraulic Fracturing Study: Chemical & Analytical Methods February 24-25, 2011, Arlington, VA, U.S. Environmental Protection Agency, 2011.

42. I. Ferrer, E.M. Thurman, Liquid Chromatography Time-of-Flight Mass Spectrometry: Principles, Tools and Applications for Accurate Mass Analysis, John Wiley and Sons, New York, 2009.

Chapter 6

Hydraulic Fracturing Stimulation Techniques and Formation Damage Mechanisms—Implications from Laboratory Testing of Tight Sandstone–Proppant Systems

Andreas Reinicke, Erik Rybacki, Sergei Stanchits, Ernst Huenges, and Georg Dresen

Helmholtz Centre Potsdam, GFZ German Research Centre for Geosciences, Telegrafenberg, 14473 Potsdam, Germany

ABSTRACT

Reservoir formation damage may seriously affect the productivity of a reservoir during various phases of fluid recovery from the subsurface. Hydraulic fracturing technology is one tool to overcome inflow impairments due to formation damage and to increase the productivity of reservoirs. However, the increase in productivity by hydraulic fracturing operations can be limited by permeability alterations adjacent to the newly created fracture face. Such an impairment of the inflow to the fracture is commonly referred to as fracture face skin (FFS).

Here, we focus on mechanically induced fracture face skin, which may result from stress-induced mechanical interactions between proppants and reservoir rock during production. In order to achieve sustainable, long-term productivity from a reservoir, it is indispensable to understand the hydraulic and mechanical interactions in rock–proppant systems.

We performed permeability measurements on tight sandstones with propped fractures under stress using two different flow cells, allowing to localise and quantify the mechanical damage at the fracture face. The laboratory experiments revealed a permeability reduction of this rock–proppant system down to 77% of initial rock permeability at 50 MPa differential stress leading to a permeability reduction in the fracture face skin zone up to a factor of 6. Considerable mechanical damage at the rock–proppant interface was already observed for stresses of about 5 MPa. Microstructure analysis identified quartz grain crushing, fines production, and pore space blocking at the fracture face causing the observed mechanically induced FFS. At higher stresses, damage and embedment of the ceramic proppants further reduces the fracture permeability. Therefore, even low differential stresses, which are expected under in-situ conditions, may considerably affect the productivity of hydraulic proppant fracturing stimulation campaigns, in particular in unconventional reservoirs where the fracture face is considerably larger compared to conventional hydraulic stimulations.

A BRIEF OVERVIEW OF HYDRAULIC FRACTURING STIMULATION TECHNIQUES

Reservoir rocks are commonly stimulated using chemical, thermal, or hydraulic techniques. The hydraulic fracturing technique allows one to considerably enlarge the accessible reservoir volume by the injection of fluids along the isolated segments of the wellbore into the formation. Theoretically, the minimum fracture initiation pressure in an elastic homogeneous and isotropic medium depends on the magnitude of the horizontal stresses, the pore pressure, and the (hydraulic) tensile strength of the rock formation (e.g.Hubbert and Willis, 1957 and Baumgartner and Rummel, 1988). Initiation of hydraulic failure is often affected by pre-existing shear zones and bedding planes, in particular if contrast between the principal stresses is small (e.g. Wallroth et al., 1996, Cornet and Julien, 1989 and Thiercelin and Roegiers, 2000).

During hydraulic stimulation, the induced fracture extends until the rate of fluid loss into the formation exceeds the pumping rate. However, once pumping ceases and pressure drops below the fracture closure pressure, the fracture may close and rapidly heal and the fractured area is no longer available for production. To avoid this, a propping agent (sand or ceramic spheres) is usually added to the fluid and deposited inside the fracture (Economides and Nolte, 2000).

Three Hydraulic Stimulation Concepts

Three different concepts exist for hydraulic stimulation of a reservoir, depending on rock, formation and fluid properties, namely: hydraulic proppant fracturing, water fracturing, and hybrid fracturing (Rushing and Sullivan, 2003).

Hydraulic Proppant Fracturing (HPF)

Highly viscous gels with high proppant concentrations are used creating highly conductive, but relatively short fractures in a permeable reservoir with a porous matrix. The fracture connects the well and the reservoir and reduces permeability impairments in the direct vicinity of the well (commonly referred to as skin (Dake, 1978)), which results in a productivity increase. After fracture generation and proppant pack emplacement are completed, the well is shut-in for some time to allow the fluid to leak off into the formation and the pressure declines. During this shut-in phase, the fracture closes partially fixing the proppant pack in place (Fig. 1).

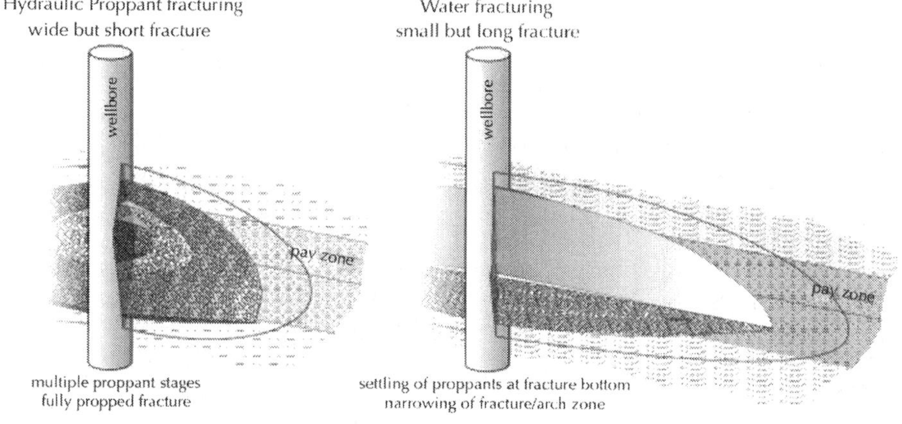

Figure 1: Hydraulic fracture generation in a reservoir. If fluid at sufficiently high pressure is pumped into a reservoir, a fracture propagates from the wellbore into the reservoir. To avoid a rapid fracture closing after leak off of the pumped fluid, usually a propping agent (sieved sand or ceramic spheres) is placed in the fracture. In dependence of the stimulation concept and the reservoir conditions, a short and wide fracture or a large and small fracture can be created.

HPF treatments are preferentially used in medium to high permeable formations (10–1000 mDa) and permit good controlling of the stimulation parameters (fracture conductivity, fracture width, fracture length, and fracture height). These parameters may be predicted and optimised. However, HPF operations may induce problems in well performance. High proppant concentration can screen out in the near wellbore region due to pressure losses. Also, the complex fluid chemistry may be incompatible with reservoir fluids resulting in precipitation of minerals. In addition, incomplete clean-up after fracturing treatment may leave behind gel residues (unbroken polymer chains) in the proppant pack. This is of particular importance in reservoirs of moderate to high permeability (Aggour and Economides, 1999) where high fracture conductivity is essential in order to achieve sufficient drainage.

Water Fracturing (WF), "Self-propped Fracs" or "Water Fracs"

Water containing friction-reducing chemicals (slick water) partially with added low proppant concentration (mainly sieved sand) is used as a pumping fluid to create long and narrow fractures (Fig. 1). A WF treatment aims at connecting reservoir parts at some distance from the borehole. In addition inflow area is maximised by connecting the well to a network of natural joints. In geothermal Hot Dry Rock applications (Baria et al., 1999), WF treatments are applied to connect two wells in a tight hard rock (e.g. granite).

Mayerhofer and Meehan (1998) compared the performance of 50 water fracturing treatments with traditional hydraulic proppant fracturing treatments in the Cotton Valley, Texas, USA. The results show that in general water fracs are at least as successful as HPF in enhancing productivity in fields with relatively low reservoir permeability. The advantage of WF compared to HPF and hybrid fracs is a considerable reduction in cost. Different studies have shown that application of WF is limited to reservoirs with permeability <1 mDa (Britt et al., 2006; Fredd et al., 2000; Mayerhofer and Meehan, 1998). In such reservoirs, it is important to maximise the inflow area because of the slow diffusion of fluid through the tight rock matrix.

The success of WF stimulation depends on the self-propping potential of the reservoir rock, i.e. on ability of the rock to maintain "unpropped fracture conductivity". The conductivity may vary by at least two orders of magnitude (Sharma et al., 2004), which depends on the residual fracture width, which is affected by the shear displacement of the fracture faces, their roughness (asperities), and by the strength of the rock (Rushing and Sullivan, 2003). However, alteration of fracture conductivity is based on parameters, which are difficult to measure or predict. In general, at production conditions the unpropped fracture conductivity is too small to support production over a significant fracture length (Britt et al., 2006). Also the newly created fracture may heal rapidly as a result of creep and pressure solution processes at the asperities, i.e. solution of minerals at regions of high stress with subsequent precipitation. Adding of propping agents (sieved sand or light weight proppants) at small concentration (~ 0.5 kg m^{-2}) to the frac fluid prevents rapid fracture closure and guarantees sufficient residual fracture conductivity at elevated fracture closure stress (>40 MPa) (Fredd et al., 2000).

The low viscosity of the injection fluids promotes proppant settling and often leads to unfavourable proppant placement, limiting the effective propped fracture length which is a key parameter of the production potential (Fig. 1). At the end of treatment, the fracture may consist of a packed bed at the bottom ("dune"), an unpropped fracture at the top and a transition zone in between ("arch" zone) (Warpinski, 2009).

Hybrid Fracturing or "Hybrid Fracs"

The term "hybrid fracs" describes various combinations of fracture stimulations using cross-linked gels, linear gels and slick water fluids. Hybrid fracturing combines the advantages of HPF and WF treatments by combining an initial slick water phase to create the fracture geometry. This is then followed by a cross-linked gel treatment that allows carrying the proppant load to the far end of the induced fracture. The geometry of the created fracture network differs from that of a conventional HPF stimulation design. For example, the fractures are considerably longer compared to HPF and the effective propped fracture length is higher (Coronado, 2007 and Rushing and Sullivan, 2003).

The polymer loading in gels is decreased reducing polymer damage. Nevertheless, the problems described with HPF treatments hold for hybrid fracs likewise.

Evolution of Fracturing Techniques for Tight Gas Reservoirs

Early attempts at stimulating and producing from formations with permeabilities in the microdarcy range date back to 1970 (Holditch and Morse, 1971). Since then, stimulation techniques were continuously improved and horizontal drilling as well as multi-stage fracturing is nowadays state of the art. Since fracture length is more important than fracture conductivity in tight gas reservoirs, water-based gel treatments became more common in the eighties and nineties (Settari and Bachmann, 2009). Later, slick water treatments with little or no proppants were successfully applied (Mayerhofer et al., 1997). The stimulated volume was gradually increased until the fracture conductivity became the productivity-limiting factor leading to the invention of the hybrid frac technique. The Bossier tight gas play, Texas, USA (Sharma et al., 2004) gives an example of the historical evolution of fracturing technology (Fig. 2).

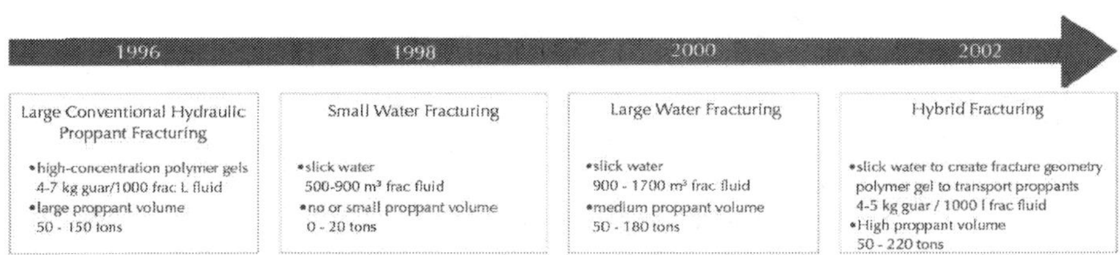

Figure 2: Evolution of fracturing technology using the example of the Bossier tight gas reservoir in Texas, USA (Rushing and Sullivan, 2003).

FORMATION AND FRACTURE CONDUCTIVITY DAMAGE MECHANISMS

Formation and fracture damage processes result from chemical, physical, biological, and thermal interactions of formation and fluids. In addition, mechanical interactions of reservoir and hydraulic fractures under stress can influence the performance. A wide range of laboratory, field, and theoretical studies cover the aspects of fracture damage mechanisms (Fredd et al., 2000, Wen et al., 2006, Behr et al., 2002, Nasr-El-Din, 2003, Moghadasi et al., 2002 and Lynn et al., 1998). A comprehensive overview of the most important damage processes in a fractured reservoir are given by Civan (2000) and Bishop (1997). The damage mechanisms in and adjacent to a propped fracture are illustrated in Fig. 3.

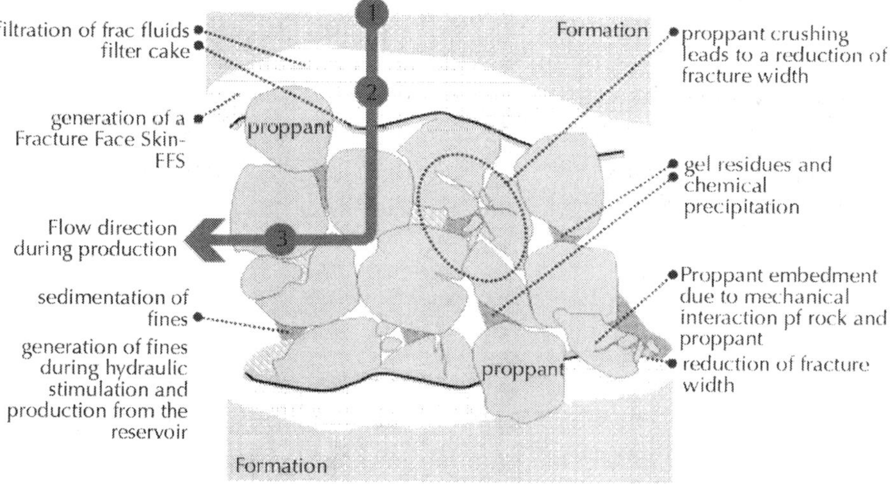

Figure 3: Mechanisms impairing proppant pack conductivity (modified from Legarth et al., 2005). Gel residues, chemical precipitation, and sedimentation of fines can affect the proppant pack permeability, as well as proppant embedment and proppant crushing in high stress environments.

A filter cake at the fracture face or filtration of frac fluids can reduce the permeability adjacent to the fracture and generate a fracture face skin (FFS). During production fluid flow from the reservoir (1) perpendicular to the fracture face (2) is affected by the FFS.

Fracture Face Skin

Cinco-Ley and Samaniego-V (1977) introduced the fracture face skin (FFS) as a potential damage mechanism. They postulated a zone with a reduced permeability at the fracture face. Any inflow from the reservoir into the (propped) fracture has to pass through this altered zone, which is commonly referred to as fracture face skin (FFS, Fig. 3). Hence, neglecting chemical effects, the fluid transport in a hydraulically fractured reservoir may be approximated by three zones with different permeabilities: (1) the reservoir rock permeability, (2) the permeability of the altered fracture face, and (3) the proppant pack permeability in the fracture (Fig. 4).

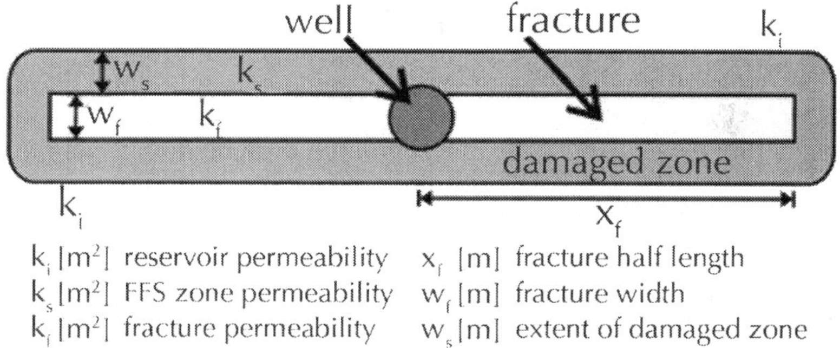

Figure 4: Schematic model of a 3-zone permeability distribution in a fractured reservoir with a fracture face skin (FFS).

Several numerical studies (Gdanski et al., 2005, Adegbola and Boney, 2002 and Behr et al., 2002), experimental work (Roodhart et al., 1988 and Ahmed et al., 1979), and site investigations (Cramer, 2005 and Lynn et al., 1998) investigated the influence of a FFS on the reservoir properties. A FFS can be caused by numerous effects like fluid-loss damage (Cinco-Ley and Samamiego-V, 1981), filter cake build-up at the fracture face (Romero and Economiedes, 2003), relative gas–water permeability changes at the fracture face (Holditch, 1979), or liquid condensates within gas condensate reservoirs (Wang et al., 2000).

During ongoing depletion of the reservoir the increasing effective stress on the propped fracture may further reduce the fracture conductivity. For example, proppant embedment and crushing may reduce the fracture width. In addition, increasing effective stress may reduce the rock permeability of tight reservoir rocks, in particular for gas transport along microfractures (e.g. in gas shales).

Influence of Fracture Face Skin in Tight Reservoirs

In a tight reservoir rock the inflow performance is considerably reduced by a FFS during early production but may gradually recover during further production (Settari and Bachmann, 2009). Upon subsequent lowering of the wellbore pressure, the frac fluid will flow back from the reservoir to the well cleaning up the fracture.

Cinco-Ley and Samamiego-V (1981) have shown that the transient well performance is mainly influenced by FFS during the period of formation of linear flow, i.e. at early times of the production. In this flow period, the pressure loss caused by flow resistance in the fracture is negligible, and the pressure transient behaviour of the well is governed by the linear flow behaviour normal to the fracture plane. At late times of production, i.e. during pseudo-radial flow period the influence of a FFS on productivity is small. The gas transport in low permeable formations («1 mDa) may be governed by linear flow over years. Hence, FFS damage is particularly important in low-permeability reservoirs.

In tight gas reservoirs, the production analysis often indicates a post-fracture well productivity that is significantly lower than expected from fracture characteristics simulations (Cramer, 2005). Field tests using water fracs with significant amounts of proppants have delivered fracture conductivities comparable to conductivities of unpropped fractures (Settari and Bachmann, 2009). Hence, there is still a need for better understanding and investigating the complex mechanisms influencing the permeability alteration of rock–proppant systems.

Mechanically Induced Fracture Face Skin

We expect a mechanically induced FFS resulting from the interaction between proppants and rock under non-isostatic stress conditions. Differential stress may cause crushing of grains and fines production at the fracture face and the embedment of proppants into the rock matrix (Fig. 3). The produced fine particles block the pores at the fracture face and this may result in a reduced permeability and an additional pressure drop normal to the fracture face.

The aim of this work was the development of new laboratory equipment to locate the damaging effects and to quantify the mechanically induced FFS in terms of permeability reduction and penetration of the damaged zone.

EXPERIMENTS

Materials and Methods

Specimens were prepared from tight Flechtingen sandstone, sampled from an outcrop in Saxony/Germany (8–10% porosity, 100–200 μDa permeability). The sample material is a Lower Permian (Rotliegend) sandstone containing 65% quartz, 13% feldspar, 9% illite, and 4% carbonates (Trautwein, 2005). Testing parameters are given in Table 1. Two kinds of proppants were used in this study: (1) intermediate strength proppants (ISP), designed for a maximum fracture closure stress of ~70 MPa and made from fused ceramics with 20/40 mesh size (proppant particle diameter: 0.4–0.8 mm) with a mean diameter of about 760 μm and (2) high strength proppants (HSP) composed of sintered bauxite with 20/40 mesh and a mean diameter is about 700 μm. These are designed for a maximum fracture closure stress of ~100 MPa.

Table 1: Testing parameters for the rock–proppant interaction (RPI) experiments using Flechtingen sandstone, with intermediate and high strength proppants (ISP and HSP)

AEFC: rock testing		**Rock–proppant interaction testing**		
	no. 1		**no. 2**	
Specimen length	120 mm	Specimen length	123.5 mm	
Specimen diameter	50 mm	Initial fracture width	3.46 mm	
Conf. pressure	10 MPa	Conf. pressure	10 MPa	
Diff. stress	0–65 MPa	Diff. stress	0–50 MPa	
Strain rate	8.30×10^{-6} s^{-1}	Strain rate	8.1×10^{-6} s^{-1}	
Flow rate	0.02 ml min^{-1}	Flow rate	0.02 ml min^{-1}	
Initial permeability	95 μDa	Proppants	10 kg m^{-2} HSP	
BDFC: rock testing		Rock–proppant interaction testing		
	no. 3		no. 4	no. 5
Specimen length	120 mm	Specimen length	63.3 mm	65.3 mm
Specimen diameter	50 mm	Initial fracture width	3.9 mm	4.8 mm
Conf. pressure	10 MPa	Conf. pressure	10 MPa	10 MPa
Diff. stress	0–65 MPa	Diff. stress	0–50 MPa	0–50 MPa
Strain rate	8.30×10^{-6} s^{-1}	Strain rate	1.57×10^{-6} s^{-1}	1.53×10^{-6} s^{-1}
Flow rate	0.05 ml min^{-1}	Flow rate	0.05 ml min^{-1}	0.05 ml min^{-1}
Initial permeability	198 μDa	Proppants	10 kg m^{-2} HSP	10 kg m^{-2} ISP

Experimental Set-Ups

Following the approach outlined in Fig. 4, we developed two different flow cells to separate the permeability evolution with increasing stress of the reservoir rock, FFS, and of the propped fracture. (1)

The acoustic emission flow cell (AEFC) was used for analysis and localisation of crushing and damage at the rock–proppant interface during loading. (2) The bi-directional flow cell (BDFC) was designed to simulate the geometric flow conditions in reservoirs intersected by a proppant filled fracture allowing quantifying the permeability reduction in the stressed rock–proppant system.

A schematic view of both flow cells is given in Fig. 5. Five laboratory rock–proppant interaction (RPI) experiments were conducted, two with the AEFC and three with the BDFC set-up (Table 1). The properties of intact rock specimens and rock–proppant systems were investigated in experiments no. 1 and 3, and experiments no. 2, 4, and 5, respectively.

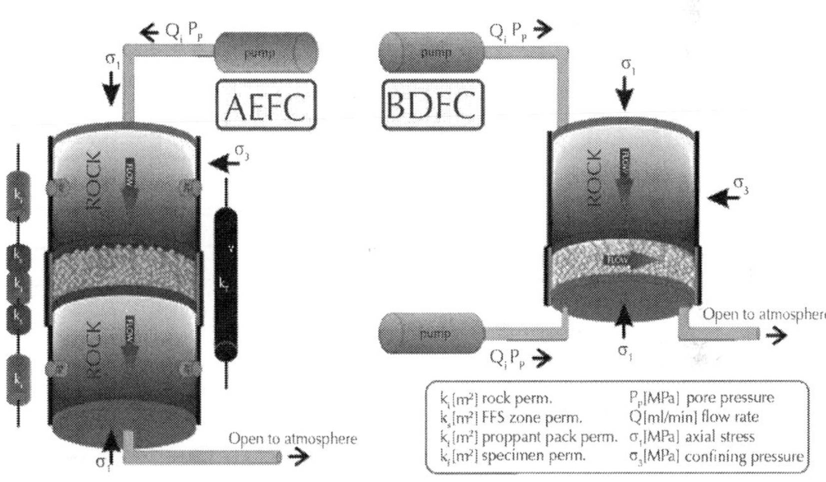

Figure 5: Experimental assemblies for measuring the permeability with rock–proppant interaction (RPI) under stress. The acoustic emission flow cell (AEFC) is equipped with 12 piezoelectric P-wave transducers, used to identify microfracturing processes at the contact between rock and proppants. The bi-directional flow cell (BDFC) allows permeability measurements for two flow directions, normal and parallel to the fracture face.

The BDFC set-up provides fluid flow in two directions, normal and parallel to the fracture face, allowing the determination of the permeability of both the rock–proppant system (overall assembly permeability including proppant pack, k_T) and of the proppant pack (k_f) solely (see Fig. 5).

The set-up is approximated as a connection in series of three permeabilties k_i, k_S, and k_f (Fig. 5). With regard to the length of each segment (L_i, L_S, and L_f) following equations for the FFS permeability k_S is derived (Reinicke, 2010):

$$k_S = \frac{k_T k_f k_i w_S}{k_i k_f L_T - k_T (k_f L_i + k_i w_f)} \tag{1}$$

The notation of the symbols is given in Table 2.

Table 2: Quantities defining the hydraulic properties of rock–proppant systems

kT	Specimen permeability	LT	Length of specimen
ki	Rock permeability	Li	Length of rock sample
kf	Proppant pack permeability	wf	Fracture width
kS	FFS zone permeability	wS	Extent of FFS zone/damage penetration

The bottom plug cylinder of the BDFC has two pressure and two flow ports. The two flow ports are connected to flow distribution circle segments to the left and right of a PEEK plate (Polyetheretherketon), which allow for a homogeneous flow field in the centre of the PEEK plate. During permeability determination of the proppant pack the differential pore pressure is measured between two small slots with a distance of 25 mm in the centre of the PEEK plate where Darcy's law is valid.

The PEEK plate is used as support for the proppant pack. The synthetic material PEEK is stiff enough to support the load, but the Young's modulus is low enough to avoid proppant crushing in contact with the PEEK (Young's modulus of 3.7 GPa/tensile strength of 90 MPa). Plastic deformation of the PEEK will embed proppants and distribute the stress. Using the AEFC allows performing experiments with fluid flow perpendicular to the fracture faces crossing the propped fracture. Twelve ultrasonic P-wave transducers (PZT) with a resonant frequency of ~1 MHz were attached to the specimen (experiment no. 2). The PZT were used to localise the acoustic emission (AE) events during grain crushing and to periodically measure P-wave velocities parallel and normal to the loading direction. Software, developed at the Helmholtz Centre Potsdam GFZ, Germany, was used for automatic picking of first motion amplitudes and for automated AE hypocenter location. First motion polarities were applied to discriminate AE source types in tensile, shear, and collapse (T-, S-, and C-) events (Zang et al., 1998). The hypocenter location algorithm is based on the downhill simplex algorithm (Nelder and Mead, 1965) modified for anisotropic and inhomogeneous velocity fields.

Experimental Procedures

A stiff (0.72×10^9 N m^{-1}) servo-controlled loading frame (MTS, Material Test Systems Corporation, Minneapolis MI, USA) was used to apply axial load, measured with a high accuracy 1000 kN load cell (sensitivity±1 kN). For AEFC experiments the confining pressure was servo-controlled with an accuracy of about 0.1 MPa. The BDFC was designed to work with a conventional, manually driven Hoek Cell (Hoek and Franklin, 1968) for application of confining pressure; the accuracy is about 1 MPa.

Experiments were done at ambient conditions. Fluid is delivered to the sample from the pore fluid system equipped with a pump (Quisix 6000). Flow direction is from top to bottom with the downstream side open to the atmosphere. Pure (demineralised) water was used as a pore fluid. The pore pressure gradient was determined using a high-resolution differential pressure transducer (Honeywell/Sensotec TJE BD121BN) with a pressure range of 3 bar and a full span error of 0.1%.

First, the initial rock permeability (k_i) was determined. Then, a tensile fracture – comparable to a hydraulic fracture – was created in the test specimen via a 3-point bending test (Sun and

Ouchterlony, 1986). Subsequently, the fracture was filled with a proppant pack (concentration: 2 lb ft^{-2}, i.e. ~10 kg m^{-2}). Initially, a constant confining pressure was applied (σ_3=10 MPa) and then specimens were loaded axially to simulate fracture closure under production conditions. Loading was stopped at defined differential stress levels (σ_{diff}=5, 20, 35, 50 MPa) and differential pore pressure (ΔP) was measured at a constant flow rate (Q) of 0.02 ml min^{-1}. At a defined stress level, the stress was kept constant for one hour and the permeability was measured according to Darcy's law (Darcy, 1856):

$$k = \eta \frac{Q}{A} \frac{\Delta l}{\Delta P} \tag{2}$$

where A is the cross-sectional area of the rock sample or proppant pack, η is the dynamic viscosity, and Δl is the measurement length.

EXPERIMENTAL RESULTS

AEFC: experiments no. 1 and no. 2: Flechtingen Sandstone+HSP

Two experiments were conducted with the AEFC. The specimens (intact rock and rock–proppant system) had a length of 120 and 123.5 mm, respectively. Specimen diameter was 50 mm; confining pressure was 10 MPa (Table 1).

Fig. 6 shows the AE hypocenters, as well as the permeabilities of these experiments. The initial permeability (k_i) of intact rock specimen was 94±2 μDa with a poroelastic permeability decrease of 10% within the applied loading range (Fig. 6, second row).

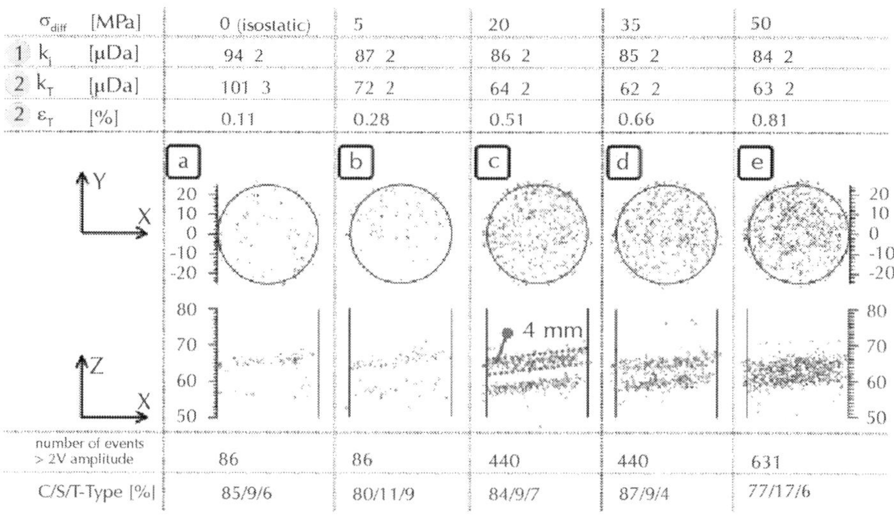

Figure 6: Permeability evolution and AE hypocenters of a tight Flechtingen sandstone–HS–proppant system. The initial permeability (k_i) was 94±2 μDa with a poroelastic permeability decrease of ~10% within the applied stress interval of 50 MPa (experiment no. 1). The permeability of the rock–proppant system (k_T) was reduced by about 30% at maximum differential stress (σ_{diff}) (experiment no. 2). The AE activity was located at the fracture faces and moved into the proppant pack with increasing differential stress. The main activity is C-type, indicating a high amount of pore collapse. (C-type: collapse event/S-type: shear event/T-type: tensile event).

At initial isostatic stress conditions, the whole assembly permeability k_T was about 100 μDa, almost similar to ki. At σ_{diff} of only 5 MPa, k_T was reduced to 72±2 μDa. At maximum differential stress, k_T amounted to 63±2 μDa (Fig. 6, third row). The majority of deformation of the rock–proppant system (ε_T=0.51%) takes places within the first two loading steps (Fig. 6, fourth row).

In the first three loading steps, the AE activity was located at the fracture face (Fig. 6a–c), i.e. all damage events were located at the contact of rock and proppants. The acoustically determined thickness of the damage zone is about 4 mm. Some 'outliers' may be attributed to the localisation accuracy of a single AE event, which is about ±1.2 mm. At increased AE activity, the overall accuracy is expected to be higher for statistical reasons.

Above 20 MPa differential stress, AE activity within the proppant pack was observed, i.e. crushing of proppants started. At maximum differential stress, many events from the proppant pack were recorded.

During the experiment, S-type events increased slightly (9–17%) and C-type events decreased (85–77%). The number of T-type events was almost constant, varying from 6% to 9%. The high amount of C-type events indicates compaction by pore collapse leading to porosity reduction.

BDFC: experiments no. 3–5: Flechtingen Sandstone+HSP and ISP

Three experiments were conducted with the BDFC. The intact specimen had a length of 120 mm; the rock–proppant system (using HSP and ISP) had a length of 63.3 mm and 65.3 mm, respectively. Specimen diameter was 50 mm; confining pressure was 10 MPa (Table 1). Before loading σ_{diff} was negative since the confining stress was higher than the axial stress. Small axial stress was chosen in order to limit fracture face damage and initial permeability reduction of the rock–proppant system.

The results of permeability measurements versus applied stress σ_{diff} of experiments no. 3–5 are shown inFig. 7. The initial permeability of this sandstone (ki) was about 200 μDa (with a poroelastic permeability change of about 15%) (exp. no. 3, Fig. 7a). At σ_{diff}=−7 MPa, which corresponds to an axial stress of 3 MPa, the permeability of the rock–proppant systems (k_T) already shows a clear reduction compared to k_i for both proppant types. At maximum σ_{diff}, the permeability is reduced to 77% (HSP) and 82% (ISP), compared to the rock permeability k_i (Fig. 7b and f). This significant reduction is most likely an effect of the mechanical interaction of rock and proppant and the accompanied destruction and fines production and pore blocking at the fracture face.

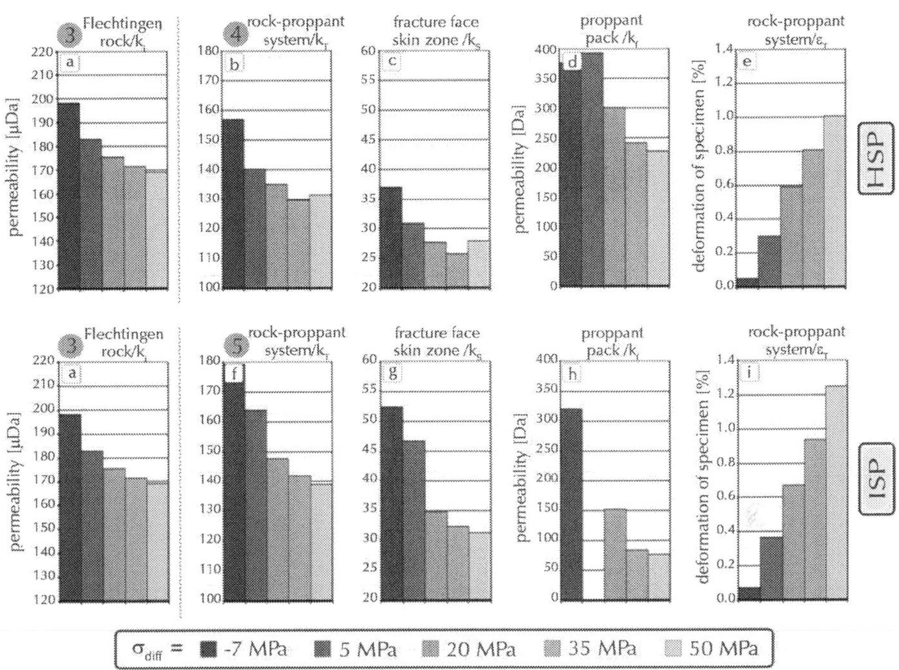

Figure 7: Permeability of rock–proppant systems as a function of differential stress (σ_{diff}). In comparison to the sandstone permeability (k_i/experiment no. 3/a), a pronounced reduction of the rock–proppant system permeability (k_T/experiments no. 4 and 5/b and f) was observed for intermediate and high strength proppants (ISP and HSP). This is effected by a significant reduction of permeability at the fracture face (c and g). The permeability of ISP shows a drastic reduction at differential stresses >20 MPa, in contrast to the HSP permeabilities (d and h). The permeability reduction of the rock–proppant system is accompanied by an increasing deformation of the specimen up to 1.2% (e and i).

The permeability of the FFS zone (k_S) is calculated using equation 1. For this purpose the extent of the FFS zone (w_S) is assumed to be 4 mm as determined from the AE experiment (see previous paragraph). Already at small differential stress, the permeability of the FFS zone (k_S) is reduced to 37±6 mDa (HSP) and 52±6 mDa (ISP), respectively (Fig. 7c and g). At σ_{diff}=50 MPa a further reduction to 28±6 mDa (HSP) and 31±6 mDa (ISP) is observed. That means compared to ki the permeability at the fracture face is reduced by a factor of 6. Fig. 7d and h shows the simultaneously measured permeability of the proppant pack (k_f). The initial values of k_f were 390±160 and 320±110 Da for HSP and ISP, respectively. The ISP permeability was reduced to 25% at maximum σ_{diff}, and the HSP permeability decreased to 58% of initial proppant pack permeability (Fig. 7c and e). HSP permeabilities are within the range of manufacturer's data (550–300 Da) with respect to the applied stress interval, whereas the ISP permeabilities show a significant discrepancy to the manufacturer's data (570–210 Da).

In accordance to the higher permeability reduction of the proppant pack, the rock–proppant system with ISP shows a higher deformation (ε_T=0.1–1.2%) within the loading range compared to the HSP (ε_T=0.05–1.0%) (Fig. 7e and i).

Microstructural Observations and Fracture Patterns

Fig. 8a shows an optical micrograph of the proppant pack and adjacent rock of experiment no. 2. At the fracture face, some quartz grains in contact with proppants were crushed and fines were produced, blocking the neighbouring pores and the flow paths. Almost no damage of HSP can be observed. Fig. 8b illustrates the proppant imprints (proppant embedment) into the fracture face and a large amount of fines produced from crushed proppants. Also, the quartz grains below the contact are crushed, indicated by white spots on the red Flechtingen sandstone.

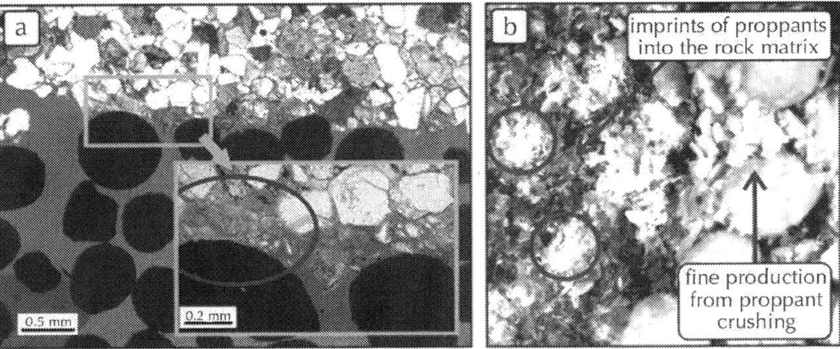

Figure 8: Micrographs of rock–proppant systems after loading. (a) At the rock–proppant (HSP) contact rock grains are crushed and fines are produced. These fines clog the neighbouring pores (inset) and produce the mechanically induced fracture face skin. (b) Fines produced from crushed ISP, which impair the proppant pack permeability.

DISCUSSION

Measurements conducted with the BDFC highlight that both proppant types create a clear reduction of permeability at the fracture face due to crack damage of proppant and rock grains (see Fig. 7). Already at small differential stresses (~20 MPa) a significant reduction of permeability (HSP: ~77%/ISP: ~84% of k_i) is observed. This leads to a reduction of permeability at the fracture face up to a factor of 6 compared to the initial rock permeability.

The AE activity at the fracture face results from embedment of proppants into the rock matrix and a resulting destruction of rock grains. The produced fines are blocking the pores at the fracture face and result in a permeability reduction of the rock–proppant system and a mechanically induced FFS. This build-up of the mechanically induced FFS is solely an effect of grain crushing at the interface.

During the first loading step (σ_{diff}: 0–5 MPa) the AE activity at the interface is relative low whereas a significant reduction of permeability is observed (Fig. 6). In contrast, in the second loading step (σ_{diff}: 5–20 MPa) a good correlation between AE events and permeability reduction is noticed. With further loading the ongoing permeability reduction is relatively low. Therefore, we assume that relatively few fines, initially produced and rearranged during the first loading, are sufficient to block the pores at the interface. In addition, the relatively low AE activity observed below 20 MPa may be an artefact of the chosen AE detection threshold of 2 V that that did not allow to record small AE events. Once the pore throats are blocked by the produced fines, further loading is expected to result in a denser crack network but to have just a minor influence on the permeability reduction, which is in accordance with the observations (Fig. 6).

For both HSP and the ISP fracture permeability is reduced due to loading the fracture up to σ_{diff}=50 MPa (~58% and ~25% of the initial k_f). This reduction is a result of three different effects: production and transport of fines from crushed rock grains, crushing of proppants and reorganisation of proppants. In standard flow cells used for proppant pack permeability measurements, the proppant pack is loaded between smoothed platens instead of rough rock surfaces (Anderson et al., 1989). The asperities of a rough surface cause small contact areas accompanied by high stress concentrations. Hence, proppant and grain crushing is promoted at smaller external stresses compared to tests with platens.

Stress Calculation in Rock–proppant Systems: Diametral Load of Proppants and Quartz Grains

An analytical approach of Hiramatsu and Oka (1966) was employed to analyse the mechanical interaction of rock grains and proppants under diametral load. For that purpose, the particles in contact were approximated as isotropic elastic spheres and a simple cubic packing of spheres was assumed. This approach allows determining the complete stress field within a sphere under uniaxial compression (diametral load). The stresses at HSP–HSP contacts (PP contact) and HSP–quartz contacts (PQ/QP contact) were computed at four loading steps (σ_{diff}=5, 20, 35, 50 MPa), comparable to the present experimental conditions. The contact radii and contact angles (Θ) at the defined loading steps were calculated using the Hertzian contact theory (Hertz, 1882), which requires knowledge of the Young's modulus and Poisson ratio of the materials in contact that are exposed to a loading force F (Fig. 9a). If the stress overcomes the tensile strength of particles, tensile splitting along the axis of compression will induce disintegration of the spheres. Hence, the location and onset of crack initiation can be estimated from the determination of the tensile stress field and its inhomogeneity along the axis of compression. The mechanical parameters used for this calculation are listed in Table 4.

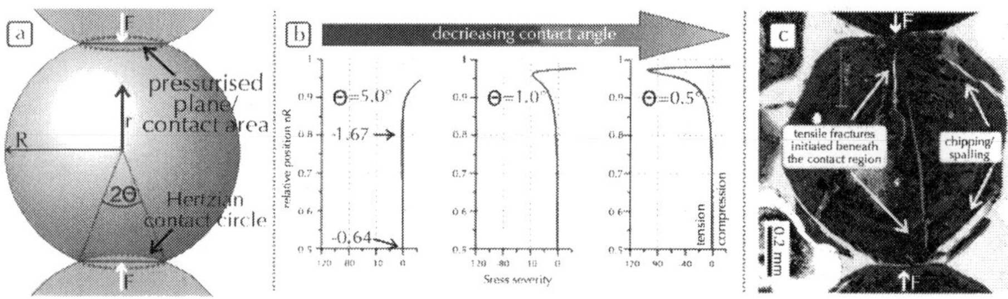

Figure 9: Stress distribution and failure type of diametrally loaded spheres (proppants/quartz grains with radius R). (a) Diametral loading geometry (R: radius; Θ: contact angle; F: loading force). (b) Decreasing contact angles induce strongly increasing tensile stress concentration beneath the surface, whereby the extremum is shifted towards the surface (r/R: 0.8, 0.95, 0.97). (c) The tensile stress concentration very likely initiated fracturing perpendicular to contact region that propagated inwards as shown in the thin section micrograph of a broken proppant. Note additional chipping of fragments at the surface.

The stress distribution along the loading axis of a corundum sphere for contact angles between 5.0° and 0.5° is given in Fig. 9b, expressed as the stress severity, which defines the ratio of the local stress with respect to the stress in the cross-sectional area of the sphere. Up to a fraction of about 0.7 of the radius R, the stress distribution within the sphere is almost uniform. However, for small contact angles (Θ=0.5 and 1.0°), a high degree of stress anisotropy exists with a severity up to ~−100 at r/R~0.95, where r is position. With increasing contact angle, the stress concentration decreases (severity of −1.67 at r/R=0.8) and the extremum is shifted towards the centre (Fig. 9b). For a low contact angle, the high tensile stress concentrations may initiate a fracture that propagates through the centre of the sphere, if the tensile stress overcomes the tensile strength. Fig. 9c shows a micrograph of a thin section prepared from a broken ceramic proppant with two fractures that are likely initiated in the contact region.

Stress maxima at grain contacts are used to calculate maximum tensile stresses in quartz grains and proppants. For this purpose, the force acting on a single quartz–proppant or proppant–proppant contact is averaged using the applied external differential stress and the number of proppants per layer (~3500). The maximum tensile stresses σ_{tPP}, σ_{tPQ} and σ_{tQP} at HSP–HSP, HSP–quartz, and quartz–HSP contacts for external stresses of 5–50 MPa are plotted in Fig. 10. The contact angles vary from 3.6° to 9.0° resulting in maximum stress concentrations (severity) between −3.3 and −0.84. For a small differential stress of 5 MPa, the maximum tensile stress in a quartz grain is about 45 MPa, which is in the range of tensile strength of quartz (60±20 MPa). Hence, the high AE activity recorded at the rock–proppant interface at small loads likely results from quartz grain crushing. This is also supported by the microstructure observations (Fig. 8a).

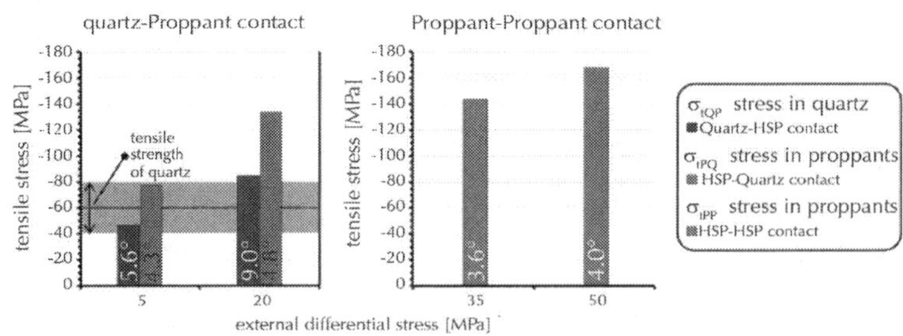

Figure 10: Maximum tensile stresses in quartz grains and proppants as a function of the external differential stress. Already at small external dif-

ferential stresses (5 MPa) the maximum internal tensile stress in quartz (σ_{tQP}) exceeds the tensile strength of quartz (blue area). In contrast, the tensile stresses in HSP at the rock–proppant interface (σ_{tPQ}) are too small to initiate failure (tensile strength of HSP: 250±50 MPa). At maximum differential stress (50 MPa), the tensile stresses at HSP–HSP contacts, i.e. within the proppant pack are smaller than the tensile strength of HSP. The numbers at the bars express the contact angels. (For interpretation of the references to color in this figure legend, the reader is referred to the web version of this article).

On the contrary, the tensile stress in proppants is far below the tensile strength 250±50 MPa for HSP–quartz contacts (green bars in Fig. 10). Even at maximum differential stress of 50 MPa, the tensile stress of ~170 MPa at HSP–HSP contacts (red bars) is below the tensile strength of corundum. Therefore, the applied analytical approach does not fully explain the failure of HSP observed at elevated differential stresses in experiment no. 2 (Fig. 6). However, the total number of events recorded from proppant pack and rock–proppant interface is small (σ_{diff}=35 MPa: 440/σ_{diff}=50 MPa: 631) compared to the total number of proppants in the fracture (~12,000). The crushing of some HSP results from small contact angles due to asperities on the proppant surfaces as well as from potential defects present in the ceramic proppants. Additionally, the approximation of the proppant pack by a simple cubic packing may oversimplify the contact geometry, the local load at a single contact in a natural system of packed spheres may be higher than calculated from this approximation.

As indicated by acoustic emission hypocenter locations, damage at the fracture face in a loaded rock–proppant system starts at relatively small nominal stresses due to significant stress concentrations at grain–proppant contacts. Consequently, microscopic local stresses in a propped fracture may differ significantly from the macroscopic nominal stresses applied to the rock–proppant system. For the materials tested in this study, strength of quartz grains from the sandstone is smaller compared to that of the ceramic proppants. In particular, in soft formations (e.g. shales) the high competence contrast between rock material and proppant may lead to substantial proppant embedment and a rapid fracture closure during reservoir depletion.

Implications for Propping Agents

The analytical stress calculation highlight that the contact angle between particles, which is a function of Young's modulus, Poisson ratio, and loading force, strongly influences the stress distribution in a sphere. A large contact angle will suppress tensile stress concentration and stabilise the proppant. Hence, modifications of material parameters (decrease of Young's modulus/increase of Poisson ratio) have the potential to reduce embedment and proppant crushing.

Weaver et al. (2005) and Penny (1987) demonstrated that sands and coated proppants with comparable grain sizes generate significantly smaller "craters" in the rock matrix compared to ceramic proppants (Table 3), i.e. proppant embedment is reduced Compared to ceramic proppants, sand has a smaller Young's modulus. Resin coating deforms plastically already at small stresses. Both mechanisms lead to a larger contact areas and reduce stresses at the rock–proppant interface, whereby the resin coating has the significantly larger effect.

Table 3: Influence of proppant properties on the size of embedment. (From Penny (1987))

Proppant type	Ceramic	Sand	Sand resin-coated
Fracture closure stress (MPa)	25	25	50
Relative size of proppant embedment			
	100%	~75%	~25%

This study is limited to uncoated proppants. Other materials such as sand, specialised very light propping agents, porous ceramics or resin-coated proppants might have a smaller damage potential, i.e. the embedment will be reduced. A resin coating act as a plastic deformable material stabilizing the proppant grains and reducing fines production in the proppant pack. With such proppant types, the AE activity at the rock–proppant interface and within the proppant pack should be significantly reduced. Such investigations would be proposed for future experiments .

Table 4: Mechanical parameters used for analytical stress calculations in quartz grains and proppants (Shackelford and Alexander, 2000)

	Quartz grain	**Proppant (corundum)**
Young's modulus (GPa)	80	380
Poisson ratio	0.17	0.23
Tensile strength (MPa)	60±20	250±50
Diameter (µm)	300	700

Implications for Hydraulic Stimulations

The effect of a small FFS layer (width of ~4 mm in our experiments, see Fig. 6) with reduced pore space and permeability on the productivity is expected to be small under field conditions. For a layer with a thickness of the order of decimetres to meters with permeability reduced by 90–99%, well productivity will be reduced significantly (Cramer, 2005, Adegbola and Boney, 2002 and Holditch, 1979). Hence, the productivity impairment due to a relatively thin layer of mechanically induced FFS could be neglected. The additional pressure drop caused by the mechanically induced FFS is very small compared to the overall pressure drop in the reservoir. However, secondary effects may be triggered due to the presence of a zone with reduced porosity at the fracture face. The reduced pore space may act as filter collecting fines from the reservoir fluid during production. Al-Abduwani et al. (2003) found in static internal filtration experiments with Bentheim sandstone that a suspension containing fine particles (pore throat radius to particle radius ratio: 12.5) leads to considerable formation damage (~90% permeability reduction). This study highlights the potential impact of a small layer with reduced porosity at the fracture face. In particular in tight reservoirs this secondary effect is emphasised by phase trapping in the zone with significantly reduced porosity. An aqueous phase pumped into the formation during fracturing generates high and irreversible water saturation in the pore throats due to high capillary forces (water blockage). The porosity available for gas flow and therefore the relative gas to water permeability is reduced. The same consideration holds for condensation of gas at the fracture face.

The initial mechanically induced permeability reduction at the fracture face may thus cause a long-term impairment of a reservoir.

Cramer (2005) identified a FFS from pressure build-up analysis of a well completed in a tight gas reservoir (permeability k~30 μDa). A numerical simulation revealed a permeability reduction of 98% in a 10–13-cm-thick layer adjacent to the fracture. About 80% drop of the wellbore pressure at early stages of production was caused by the fracture face skin.

Another possible source of FFS formation is gas condensation due to abrupt changes in pT-conditions that can form a "bank" of condensate at the fracture face. This bank can extend for several meters and severely reduce the productivity of a reservoir (Butula et al., 2005 and Dehane et al., 2000).

CONCLUSIONS

The rock–proppant interaction (RPI) experiments with the AEFC point out that grain crushing, proppant embedment, and fines production start at low differential stresses (~5 MPa) at the fracture face. With increasing differential stress (>20 MPa), the AE activity shifts into the proppant pack indicating increasing proppant failure. The analytical stress modelling identified high tensile stress concentrations at the rock–proppant interface responsible for early onset of acoustic emission and quartz grain crushing at low nominal stress.

At high differential stress, the permeability at the fracture face is reduced by a factor up to 6 compared to the initial rock permeability. The strength of proppants (ISP or HSP) has only a small influence on the permeability reduction at the fracture face. The ISP have a slightly smaller damage potential and the HSP maintain a significantly higher proppant pack permeability at elevated stresses. Optical investigations reveal abundant fines produced from crushed grains at the rock–proppant interface. Porosity at the fracture face is significantly reduced by blocking the pore space with fines. These observations indicate the evolution of a mechanically induced FFS for rock–proppant systems exposed to non-hydrostatic stresses.

The direct effect of the mechanically induced FFS on the well productivity at field conditions is expected to be negligible. In comparison, effects like fluid-loss into the fracture face, filter cake build-up at the fracture face or relative permeability changes may generate a significantly larger FFS. However, secondary effects may significantly reduce the reservoir productivity, in particular in unconventional reservoirs.

The presented investigations highlight the importance of understanding the complex mechanical-hydraulic coupled effects in rock–proppant systems. Laboratory investigations of rock–proppant systems supplement the knowledge gained from field stimulations and help to define a best practice.

ACKNOWLEDGMENTS

The authors express their gratitude to Prof. Dr. P.L.J. Zitha and Dr. R. Jung whose comments helped to improve the manuscript considerably.

Funding by BMBF (Grant 03G0671A/B/C) is gratefully acknowledged.

REFERENCES

1. Adegbola, K., Boney, C., 2002. Effect of fracture face damage on well productivity. SPE 73759.
2. Aggour, T.M., Economides, M.J., 1999. Impact of fluid selection on highpermeability fracturing. SPE 36902.
3. Ahmed, U., Abou-Sayed, A.S., Jones, A.H., 1979. Experimental evaluation of fracturing fluid interaction with tight reservoir rocks and propped fractures. SPE 7922.
4. Al-Abduwani, F.A.H., Shirzadi, A., van den Broek, W.M.G.T., Currie, P.K., 2003. Formation damage vs. solid particles deposition profile during laboratory simulated PWRI. SPE 82235.
5. Anderson, R.W., Cooke, C.E., Wendorff, C.L., 1989. Propping agents and fracture conductivity in recent advances in hydraulic fracturing. SPE Monograph Series 12, 109–130.
6. Bariaa, R., Baumgartner, J., Gerard, A., Jung, R., Garnish, J., 1999. European HDR research programme at Soultz-sous-Forets (France) 1987–1996. Geothermics 28, 655–669.
7. Baumgartner, J., Rummel, F., 1988. Experience with hydraulic fracturing as a stress measuring technique in jointed rock mass. In: Proceedings of the Second International Workshop on Hydraulic Fracturing Stress Measurements, University of Wisconsin-Madison, pp. 168–204.
8. Behr, A., Mtchedlishvili, G., Friedel, T., Haefner, F., 2002. Consideration of damage zone in tight gas reservoir model with hydraulically fractured well. SPE 82298.
9. Bishop, S.R., 1997. The experimental investigation of formation damage due to the induced flocculation of clays within a sandstone pore structure by a high salinity brine. SPE 38156.
10. Britt, L.K., Smith, M.B., Haddad, Z., Lawrence, P., Chipperfield, S., Hellmann, T., 2006. Water-fracs: we do need proppant after all. SPE 102227.

11. Butula, K.K., Maniere, J., Shandrrygin, A., Rudenko, D., 2005. Analysis of production enhancement related to optimization of propped hydraulic fracturing in gazprom's yamburskoe arctic gas condensate field, Russia. SPE 94727.
12. Cinco-Ley, H., Samaniego-V F., 1977. Effect of wellbore storage and damage on the transient pressure behaviour of vertically fractured wells. SPE 6752.
13. Cinco-Ley, H., Samaniego-V F., 1981. Transient pressure analysis: finite conductivity fracture case versus damaged fracture case. SPE 10179.
14. Civan, F., 2000. Reservoir Formation Damage. Gulf Publishing Company, Houston, TX, USA.
15. Cornet, F.H., Julien, P., 1989. Stress determination from hydraulic test and focal mechanisms of induced seismicity. Int. J. Rock Mech. Min. Sci. Geomech. Abstr. 26 (4), 235–248.
16. Coronado, J.A., 2007. Success of hybrid fracs in the basin. SPE 106758.
17. Cramer, D.D., 2005. Fracture skin: a primary cause of stimulation ineffectiveness in gas wells. SPE 96869.
18. Dake, L.P., 1978. Fundamentals of Reservoir Engineering. Elsevier, Amsterdam.
19. Darcy, H., 1856. Les Fontaines Publiques de la ville de dijon. Dalmont, Paris The famous Appendix - Note D appears here.:647 & atlas.
20. Dehane, A., Tiab, D., Osisanya, S.O., 2000. Comparison of the performance of vertical and horizontal wells in gas-condensate reservoirs. SPE 63164.
21. Economides, M.J., Nolte, K.G., 2000. Reservoir Stimulation 3rd ed Wiley and Sons Ltd., United Kingdom.
22. Fredd, C.N., McConnell, S.B., Boney, C.L., England, K.W., 2000. Experimental study of hydraulic fracture conductivity demonstrates the benefits of using proppants. SPE 60326.
23. Gdanski, R., Weaver, J.; Slabaugh, B., Walters, H., Parker, M., 2005. Fracture face

24. damage–it matters. SPE 94649.
25. Hertz, H., 1882. Ueber die beruehrung fester elastischer koerper. J. Reine Angew. Math. 92, 156.
26. Hiramatsu, Y., Oka, Y., 1966. Determination of the tensile strength of rock by a compression test of an irregular test piece. Int. J. Rock Mech. Min. Sci 3, 89–99.
27. Hoek, E., Franklin, J.A., 1968. Simple triaxial cell for field or laboratory testing of rock. Trans. Inst. Min. Metall. 77, A22 (Section A).
28. Holditch, S.A., Morse, R.A., 1971. Large fracture treatments may unlock tight
29. reservoirs. Oil Gas J..
30. Holditch, S.A., 1979. Factors affecting water blocking and gas flow from hydraulically fractured gas wells. J. Petrol. Technol., 1515–1524.
31. Hubbert, M.K., Willis, D.G., 1957. Mechanics of hydraulic fracturing. Trans. AIME 210, 153–166.
32. Legarth, B., Raab, S., Huenges, E., 2005. Mechanical interactions between proppants and rock and their effect on hydraulic fracture performance. DGMK-Tagungsbericht 2005-1, Fachbereich Aufsuchung und Gewinnung, 28–29 April 2005, Celle, Deutschland, pp. 275-288.
33. Lynn, J.D., Hisham, A., Nasr-El-Din, H.A., 1998. Evaluation of formation damage due to frac stimulation of a Saudi Arabian clastic reservoir. J. Petrol. Sci. Eng. 21, 179–201.
34. Mayerhofer, M.J., Meehan, D.N., 1998. Waterfracs—results from 50 cotton valley
35. wells. SPE 49104.
36. Mayerhofer, M.J., Richardson, M.F., Walker Jr., R.N., Meehan, D.N., Oehler, M.W., Browing Jr., R.R., 1997. Proppants? We don't need no proppants. SPE 38611.
37. Moghadasi, J., Jamialahmadi, M., Muller-Steinhagen, H., Sharif, A., Izadpanah, M.R., ¨ 2002. Formation damage in iranian oil fields. SPE 73781.

38. Nasr-El-Din, H.A., 2003. New mechanisms of formation damage: lab studies and case histories. SPE 82253.
39. Nelder, J.A., Mead, R., 1965. A simplex method for function minimization. Comput. J. 7, 308–313.
40. Penny, G.S., 1987. An evaluation of the effects of environmental conditions and fracturing fluids upon long term conductivity of proppants. SPE 16900.
41. Reinicke, A., 2010. Mechanical and hydraulic aspects of rock–proppant systems —laboratory experiments and modelling approaches. Doctoral Thesis, Universitat Potsdam, Germany. ..
42. Romero, D.J., Valko´ , P.P., Economides, M.J., 2003. Optimization of the productivity index and the fracture geometry of a stimulated well with fracture face and choke skins, SPE 81908.
43. Roodhart, L.P., Kulper, T.O.H., Davies, D.R., 1988. Proppant-pack and formation impairment during gas-well hydraulic fracturing. SPE 15629.
44. Rushing, J.A., Sullivan, R.B., 2003. Evaluation of hybrid water-frac stimulation technology in the bossier tight gas sand play. SPE 84394.
45. Shackelford, J.F., Alexander, W., 2000. The CRC materials science and engineering handbook, CRC Press.
46. Settari, A., Bachmann, R.C., 2009. Reservoir and fracturing engineering challenges in tight gas development. First Break 27.
47. Sharma, M.M., Gadde, P.B., Sullivan, R., Sigal, R., Fielder, R., Copeland, D., Griffin, L., Weijers, L., 2004. Slick water and hybrid fracs in the bossier: some lessons learnt. SPE Paper 89976.
48. Sun, Z.Q., Quchterlony, F., 1986. Fracture toughness of stripa granite cores. Int. J. Rock Mech. Min. Sci. Geomech. Abstr. 23 (6), 399–409.

49. Thiercelin, M.C., Roegiers, J.-C., 2000. Formation characterization: rock mechanicsIn: Economides, M.J., Nolte, K.G. (Eds.), Reservoir Stimulation 3rd ed Wiley and Sons Ltd., United Kingdom.
50. Trautwein, U., 2005. Poroelastische verformung und petrophysikalische eigenschaften von rotliegend sandsteinen. Ph.D. Thesis, Technische Universitat, Berlin. ¨
51. Wallroth, T., Jupe, A.J., Jones, R.H., 1996. Characterisation of a fractured reservoir using microearthquakes induced by hydraulic injections. Mar. Petrol. Geol. 13, 447–455.
52. Wang, X., Indriati, S., Valko, P.P., Economides, M.J., 2000. Production impairment and purpose-built design of hydraulic fractures in gas-condensate reservoirs. SPE 64749.
53. Warpinski, N.R., 2009. Stress amplification and arch dimension in proppant beds deposited by waterfracs. SPE 119350.
54. Weaver, J.D., Nguyen, P.D., Parker, M.A., van Batenburg, D., 2005. Sustaining fracture conductivity. SPE 94666.
55. Wen, Q., Zhang, S., Wang, L., Liu, Y., Li, X., 2006. The effect of proppant embedment upon the long-term conductivity of fractures. J. Petrol. Sci. Eng. 55, 221–227.
56. Zang, A., Wagner, F.C., Stanchits, S., Dresen, G., Andresen, R., Haidekker, M.A., 1998. Source analysis of acoustic emissions in Aue granite cores under symmetric and asymmetric compressive loads. Geophys. J. Int. 135.

Citations

CHAPTER 1

Stefano Secchi and Bernhard A Schrefler, Hydraulic fracturing and its peculiarities, doi: 10.1186/2196-1166-1-8.

CHAPTER 2

Katrin Breede, Khatia Dzebisashvili, Xiaolei Liu, and Gioia Falcone, A Systematic Review of Enhanced (or Engineered) Geothermal Systems: Past, Present and Future, doi: 10.1186/2195-9706-1-4.

CHAPTER 3

Axel Bergmann, Frank-Andreas Weber, H Georg Meiners, and Frank Müller, Potential Water-Related Environmental Risks of Hydraulic Fracturing Employed in Exploration and Exploitation of Unconventional Natural Gas Reservoirs in Germany, doi: 10.1186/2190-4715-26-10.

CHAPTER 4

William T. Stringfellow, Jeremy K. Domen, Mary Kay Camarillo, Whitney L. Sandelin, Sharon Borglin, Physical, chemical, and biological characteristics of compounds used in hydraulic fracturing, Journal of Hazardous Materials, Volume 275, 30 June 2014, Pages 37-54, ISSN 0304-3894, http://dx.doi.org/10.1016/j.jhazmat.2014.04.040.

CHAPTER 5

Imma Ferrer, E. Michael Thurman, Chemical constituents and analytical approaches for hydraulic fracturing waters, Trends in Environmental Analytical Chemistry, Volume 5, February 2015, Pages 18-25, ISSN 2214-1588, http://dx.doi.org/10.1016/j.teac.2015.01.003.

CHAPTER 6

Andreas Reinicke, Erik Rybacki, Sergei Stanchits, Ernst Huenges, Georg Dresen, Hydraulic fracturing stimulation techniques and formation damage mechanisms—Implications from laboratory testing of tight sandstone–proppant systems, Chemie der Erde - Geochemistry, Volume 70, Supplement 3, August 2010, Pages 107-117, ISSN 0009-2819,

Index

M

N

P

R

S

T

U

V

W